CLAIRE LEE CHENNAULT

American Warrior

U.S.
U.S.

CLAIRE LEE CHENNAULT

AMERICAN WARRIOR

BOOK TWO

CHENNAULT TRILOGY

Lt. Col. Richard P. Voorhies Jr.
USAFR, Retired

FIRST U.S. EDITIONS
Hardcover ISBN-13: 978–1–945333–30–9
Paperback ISBN-13: 978–1–945333–31–6
Library of Congress Control Number: 2023924700

Front cover: *Chennault* by We Zudou, courtesy of the People's Republic of China through the work of Cynthia Chennault

> Presented to Anna Chennault by We Zudou in the fall of 2015 during a celebration of the seventieth anniversary of victory in the Sino-Japanese War. The painting's background shows the Karst mountains of Guilin, where the General had temporary headquarters in a cave during the war. On behalf of the late General Chennault, Anna accepted a medal from President Xi Jinping during the same event in the Great Hall of the People.

Back cover: *Richard's Tiger* by Emily Andras, 2018
Illustration credits: Maps and photos by Christopher Bull and Barbara Evenson

Available from amazon.com and other online book stores

Signed copies and volume purchases available from the author at Richard's Books, (504) 373–6665 or rvoorhiesj@cox.net

Published by Richard's Books, New Orleans, Louisiana

Printed in the United States

Dedication

For all the POW/MIAs
You are not forgotten

To the US Air Force Air Rescue & Recovery Service
That Others May Live

Nell Thompson Chennault

Brigadier General Wiltz Segura, USAF (ret.)
Perry Segura

Contents

Part Three: With the US Army Air Forces

List of Figures

Map Symbols

Rivers and oceans

Roads

Railroads

Cities

Country borders

Japanese Military

Army advance

Naval invasion

Naval blockade

Air Force attack

Air Force movement

CAF/AVG/CATF/14th AF

Air Force attack

Air Force movement

British

RAF defense

Allied Supply Routes

OPEN ground route

CLOSED ground route

HUMP air route

List of Maps

All maps adapted from US Army maps, except Map 2-1, credit: Frank Ramspott

Common Abbreviations

IJA	Imperial Japanese Army
IJN	Imperial Japanese Navy
IJAAF	Imperial Japanese Army Air Force
AVG	American Volunteer Group (Flying Tigers)
CNAC	Chinese National Aviation Corporation
USAAC or AAC	US Army Air Corps
USAAF or AAF	United States Army Air Forces (June 1941)
USAF	United States Air Force (September 1947)
CATF	China Air Task Force
CACW	Chinese American Composite Wing
CAF	Chinese Air Force
RAF	Royal Air Force (British)
WWI	World War I
WWII	World War II

Foreword

After reading *Chennault: American Warrior* by Lt. Col. Richard Voorhies Jr., I was very impressed. The work is well researched, meticulously documented and beautifully written. Lt. Col. Voorhies has masterfully composed a text that is both factual and easily readable by people not versed in the story of my grandfather, Lt. Gen. Claire Lee Chennault.

Much of the information is from previously unknown sources. This new information introduced adds depth and character to General Chennault. Voorhies covers the period of his life from the formation of the famous "Flying Tigers" and carefully follows events until the end of the Second World War.

The book deals with the unknown fight occurring in China, one of the most important theaters of WWII. All of the significant events of General Chennault's campaigns are thoroughly explained and much new anecdotal material is added in the text.

Voorhies captures the spirit and feelings of General Chennault. The how and why my grandfather did what he did has not been seen in other works. While reading the text you hear the words of my grandfather and get a feel for his thoughts, logic and actions. The book takes you on a journey and makes you feel like you are an active participant in the action. I found the text extremely informative about General Chennault's life through one of through the most difficult and trying times. Voorhies shows his failures and triumphs as he works his way through the war.

This is one of the most comprehensive and true representations of my grandfather that I have ever read. Well done, Col. Voorhies, on an exceptional book. I eagerly look to reading forward to Book 3 of the Chennault Trilogy.

Nell Chennault Calloway, Director
Chennault Military Museum, Monroe, Louisiana

Acknowledgments

To all the men and women who served with me during my military career, the lessons you taught made me a better officer and person. Because of you, I was able to accomplish all of my duties as an Air Force officer. Thank you for your guidance and help.

To my family, especially my wife, Mary, and my children, Richard and Sarah, continuing appreciation for all you have done for me over the years. To Richard, Evangeline, Grant, Jack and Marian, my grandchildren, thanks for the joy you give me.

This book would not have been possible without the support and positive reinforcement my family gave over the course of this project.

Preface

Claire Lee Chennault found himself in a unique position at the beginning of World War II. He was exactly in the right place at the right time, with a talented fighting force he had personally trained.

As events unfolded on the morning of December 8, 1941 (Asian time) he was the only American commander prepared to do battle with Japan.

Originally, Chennault intended to attack the Japanese home islands by the end of December 1941. Had he been able to launch his air campaign as planned, the Japanese could not have launched their attacks against American forces in the Pacific. This would have changed the entire course of the war in Asia.

As it was, the only successful Allied combat leader at the start of the war was Claire Lee Chennault and his small, ragtag band of American mercenaries in China and Burma.

This force, against all odds, succeeded in holding off the Japanese military long enough to allow the Allies to reorganize their forces and begin the long fight back to victory.

Richard Voorhies Jr
New Orleans, Louisiana
December 1, 2023

It is my fondest hope that the sign of the Flying Tiger will remain aloft just as long as needed and that it will always be remembered on both shores of the Pacific as the symbol of two great peoples working toward a common goal in war and peace.

Claire Lee Chennault, Lt. General USAF (Ret.)

Introduction

Chennault's difficult personality dominated his adult life, and its origins were apparent in his childhood. He faced many challenges in his personal and professional life that refined and strengthened his character. I found several ideas and traits woven through his life which explain much of Chennault's sometimes puzzling behavior as well as his great successes.

From the American South

Chennault was a product of his time and circumstances. Born in the rural south just 28 years after the American Civil War ended, he felt the effects of that war. Lessons he learned about the South's defeat carried forward into victory.

Contradictions

Deep contradictions defined much of Chennault's life. He was a rough-and-tumble boy filled with physical confidence, yet he carried insecurities forward through college and into his military career. Even as a child, Chennault was complex—bold but shy, intelligent but insecure, a young boy self-educated, rustic and rugged. He was quick to master any subject that appealed to him. Pride and assertiveness about his knowledge conflicted with insecurity regarding his academic prowess. He was a born leader but a terrible follower.

Competitiveness

Chennault drove himself hard. Second place did not exist for him. He had to be the best athlete, student, and military officer. These self-imposed demands sometimes made him inflexible and cantankerous.

Dogged Determination

Although not from the usual social class for high-ranking officers, Chennault nevertheless pursued a military career. He was not a great orator, but he vociferously advocated for his ideas. First as a civilian and later as a military officer, he fought against theories with which he disagreed. After a disappointing U.S. military career, he took a second run at proving his ideas in China, rocketing quickly to global fame and top military leadership.

Isolated nature

Chennault's nature kept him aloof from many of his contemporaries. He was uncomfortable in more formal social situations and avoided them if possible. Unbending belief in his own judgment and his rigid self-reliance proved a double-edged sword, as much a curse as a blessing. He was most comfortable working alone, often staying late into the night, whether improving classroom presentations in a one-room schoolhouse or refining his tactical theories on pursuit aviation in the military.

Love of flying

From the time of his first flight, Chennault maintained an ongoing love affair with aviation that continued throughout his life. He applied repeatedly for pilot training despite many rejections. He found a way around regulations to get flying training, anyway. Finally accepted into flight school and after a questionable start, he became a premier pilot and flight instructor.

He founded and led the first Army aerobatics troupe, Three Men on a Flying Trapeze, which became famous for its daring, seemingly impossible stunts. Then when most men may have relished retirement, Chennault took on a second career in China, because he wanted to keep on flying.

* * *

Delving into his mindset, his personality and his motivation, this book examines Chennault's development, exploring how he became the man he was and how he remained true to his beliefs. Self-reliant and self-assured, he believed himself to be correct in nearly all things at nearly all times. By often refusing to back down, he alienated himself from his peers and seemed insubordinate to his Army superiors.

Young Chennault's actions and traits cost him a successful career in the U.S. military. But failure in the U.S. led to success in China, where his qualities would flourish. How did he ascend from obscurity to greatness during the Sino-Japanese War and World War II?

Chennault's one consuming passion was flying. In this, he first failed to find recognition and satisfaction, but he carried on toward redemption and fame. On such things, a man's life, history and the survival of the world sometimes turn.

Part One

With the Chinese Air Force

1

CHENNAULT SOLDIERS ON

Preamble, summarized from *Chennault: A Rebel in China*

Chennault's career as a fighter pilot in the US Army Air Force was one of increasing responsibility and success. However, Army top brass embraced the idea that large modern bombers were the wave of the future so there was no role for defensive pursuit aircraft or pilots. In arguing vociferously against this theory and demonstrating its fallacy, he developed a reputation as an incorrigible rebel. Eventually, he retired and accepted an offer from China to inspect their air forces.

Ever since Chennault had arrived in China in June 1937, events had moved at a breakneck pace. First, he met with the President of China Chiang Kai-shek, and his influential wife, Madame Chiang. Then he flew on an extensive tour of the airbases and training fields of the Chinese Air Force, and found it woefully unprepared for war. During the inspection tour, the Marco Polo Bridge Incident happened on July 7, 1937 and that marked the outbreak of hostilities with Japan.

Chennault immediately offered his services to Chiang Kai-shek and, as an advisor, organized the Chinese air defenses. They fought valiantly and much more effectively than the Japanese had expected. However, the superior experience, equipment and numbers of the Japanese Army and Navy flyers eventually overwhelmed the Chinese

Air Force, and forced the Chinese armies to retreat. The Japanese captured Shanghai in November and Nanking in December 1937. In Nanking, their army went wild and massacred tens of thousands civilians. The remains of the Chinese Army and Air Force retreated up the Yangtze to Hankow. Only the timely arrival of Russian pilots and aircraft helped the Chinese army slow the Japanese advance up the Yangtze River.

Chennault needed aircraft. William Pawley's aircraft company CAMCO completed assembly of thirty long range Vultee V-11 bomber aircraft. Madame suggested setting up a bomber squadron using foreign pilots to lead and supplement the Chinese pilots then available in China. Ultimately, this International Squadron failed because of inadequate training, lax discipline and arrogant leadership, and led to an anti-foreigner backlash (except Russian). Gen. Mow (who spoke Russian) became the Chinese advisor to the Russian Air Force there in China. But this left Chennault without an obvious way to help China.

Chennault considers how to help China now

By February 1938, Chennault was the chief advisor to an air force that barely existed. Madame asked Chennault to take charge of all training for new Chinese pilots. However, then she was forced to resign as head of the Aeronautical Commission, being replaced by her brother, Dr. T.V. Soong, leaving Chennault in limbo. For some months, Dr. Soong waited while the political situation settled and debated whether a foreigner should be in charge of training.

Chennault had a tough decision to make. He considered returning to America, where he might return to active duty and use what he had learned fighting the Japanese.

> It seemed to me that this experience could be applied faster and more fruitfully in training American airmen than in trying to hatch a Chinese phoenix in the mountain-ringed hinterlands. The Air Corps replied officially that, for 1938, there were "no funds available for the return of retired officers to active duty." (Chennault, *Way of a Fighter*, 73)

The Air Corps needed him much more than they knew, but he had established quite a reputation as a rebel. The Chinese knew they needed him, and China fascinated him. Early in July, Dr. Soong met with Chennault and ordered him to Kunming to take charge of training. Chennault stayed and did whatever China required.

> I came to Kunming in the summer of 1938 as an exile from the air combat then raging over the middle Yangtze valley. I was spurred by direct orders from Madame Chiang to begin the seemingly hopeless task of forging a new Chinese Air Force from an American mold. (Chennault, *Way of a Fighter*, 73)

Dr. Soong gave Chennault a month to plan what should be done, and met with him again in August 1938. In addition to Training pilots, Chennault proposed plans in three other areas: Building airfields; Obtaining aircraft; and Setting up a warning network. Dr.

FIGURE 1–1. CHINESE WORKERS PULL GIANT ROLLER TO PACK GRAVEL RUNWAYS. Entire villages turned out to build a runway. They leveled the ground using hand tools. After breaking rocks into gravel, they carried baskets of gravel to lay into layers several feet thick. Then teams would pull heavy rollers over the gravel to pack and smooth the new runway area. (WARFARE HISTORY NETWORK, MATTERHORN)

Soong approved these plans and Chennault went to work rebuilding the Chinese Air Force.

Building airfields

The airfields in southern and western China were sod fields built for the biplane fighters of the 1920s. They lacked runways, because planes then had to land heading into the wind, whichever way that blew. These fields became sodden quagmires in the monsoon season from May to September. Modern fighters and bombers landed faster, so wind was less an issue, but they required longer, smoother, all-weather runways.

Thousands of Chinese broke rocks into gravel and carried that to the leveled ground. More pulled giant rollers by hand to pack the gravel into a runway.

> All over Free China these human ant heaps rose to turn mud, rock, lime, and sweat into 5,000-foot runways to nest planes not yet built in Los Angeles and Buffalo factories. Even I had a hard time keeping track of this vast effort. (Chennault, *Way of a Fighter*, 81)

Chennault envisioned building a network of bases with gravel runways suitable for bombers at the major cities in unoccupied China.

> Although their air force had virtually vanished, the Chinese began an extensive airfield-building program during the winter of 1938–39. It was during this period that the key eastern fields of Hengyang, Lingling, Kweilin, and Liuchow were modernized and equipped with stores of gas, bombs, and ammunition. Construction of Paoching and Chihkiang fields was also begun. I am proud of the part I played in building Chihkiang, for it was this field that, during 1944–45, stood fast as our last bastion in East China, defying every Japanese effort to capture it. I came to Chihkiang in the fall of 1938 when the only other foreigners in the area were a Dutch and a Belgian priest. Their mission was too far from the airfield site, which I personally selected, so I lived for two weeks with the Chinese coolies who began leveling hills for the landing area. (Chennault, *Way of a Fighter*, 81)

Rebuilding training bases

Advanced training and Headquarters moved to Kunming. Aviation gas and parts could be supplied through the neutral port of Haiphong in French Indochina via a narrow-gauge rail line to Kunming, as well as via the Burma Road, thus bypassing the Japanese blockade. Early training started in Liuchow, which had good road access to supplies from nearby French Indochina. Liuchow has been the main base of a warlord in the 1920s, isolated in the hinterlands.

However, in September 1938, the Japanese began carrier operations near Canton to prepare for invasion there. The training base at Liuchow proved vulnerable to Japanese attack. They damaged or destroyed many of the Fleet trainers in two raids on September 13 and 18.

Intermediate training then moved farther inland to Mengtze, still on the rail line to Hanoi. Primary training moved even further west to Yunnan-Yi, on the Burma Road past Kunming.

Obtaining aircraft

To start over, the Chinese Air Force had to buy new aircraft, trainers, fighters and bombers. They had already ordered 30 Hawk 75 fighters from Curtiss-Wright, but needed many more fighters and trainers. The Chinese Commission on Aeronautical Affairs under Dr. T.V Soong started this process early in 1939, asking Chennault for his recommendations. For fighter aircraft, he selected the P-35 made by the Seversky company and represented by Allen Patterson.

William Pawley was the Curtiss-Wright representative and opposed this deal since he wouldn't make any money on it. Pawley worked closely with H.H. Kung, Vice-President and Finance Minister. Initially, the Commission approved the order, but later after months of Pawley influence, rescinded it. Chennault and Soong fought with Pawley and Kung all spring and summer of 1939. In the end, Chennault did not get the P-35 order and the Seversky company went bankrupt as a result. The delay was costly for China and all involved.

What remained in the Patterson contract ashes was an order to North American Aviation for training aircraft. Harvey Greenlaw, the company representative, had a factory in Hengyang to assemble them. Greenlaw's factory was idle before the trainers could be delivered, so Pawley started assembling his Hawk 75s there. With the fall of Hankow and Canton imminent, Greenlaw moved his operation to Hanoi in neutral Indochina, and when that didn't work out in February of 1940, he and his wife went on to Rangoon. His trainers, still crated, followed behind. He could not obtain permission to

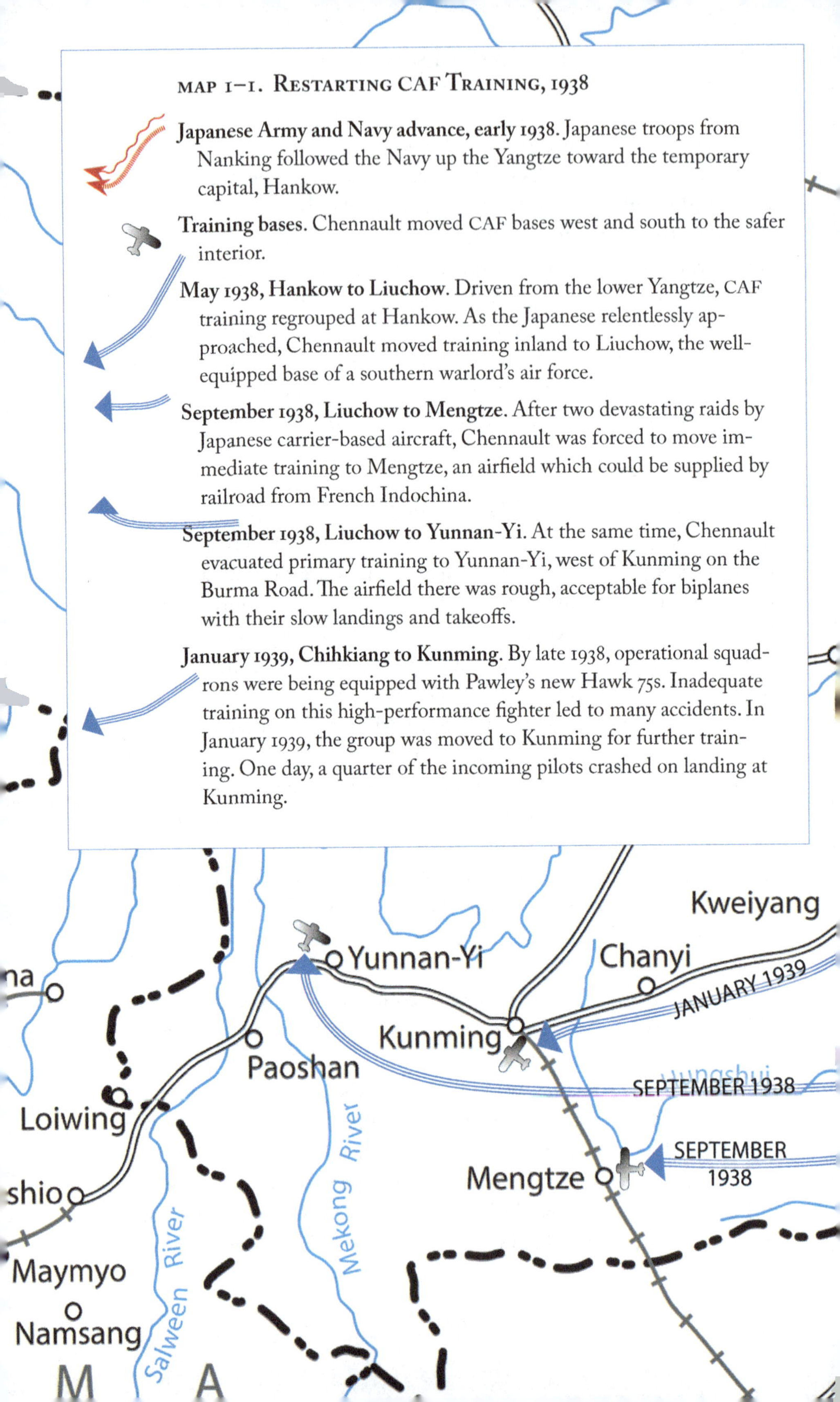
MAP 1–1. Restarting CAF Training, 1938
Japanese Army and Navy advance, early 1938. Japanese troops from Nanking followed the Navy up the Yangtze toward the temporary capital, Hankow.
Training bases. Chennault moved CAF bases west and south to the safer interior.
May 1938, Hankow to Liuchow. Driven from the lower Yangtze, CAF training regrouped at Hankow. As the Japanese relentlessly approached, Chennault moved training inland to Liuchow, the well-equipped base of a southern warlord's air force.
September 1938, Liuchow to Mengtze. After two devastating raids by Japanese carrier-based aircraft, Chennault was forced to move immediate training to Mengtze, an airfield which could be supplied by railroad from French Indochina.
September 1938, Liuchow to Yunnan-Yi. At the same time, Chennault evacuated primary training to Yunnan-Yi, west of Kunming on the Burma Road. The airfield there was rough, acceptable for biplanes with their slow landings and takeoffs.
January 1939, Chihkiang to Kunming. By late 1938, operational squadrons were being equipped with Pawley's new Hawk 75s. Inadequate training on this high-performance fighter led to many accidents. In January 1939, the group was moved to Kunming for further training. One day, a quarter of the incoming pilots crashed on landing at Kunming.
Kweiyang
Yunnan-Yi
Chanyi
JANUARY 1939
Kunming
Paoshan
Loiwing
SEPTEMBER 1938
SEPTEMBER 1938
Mengtze
Mekong River
Salween River
Maymyo
Namsang

Taierchwang
Chengchow
Su
Yellow River
Sian
N A
ang
Laohokow
Nan
JAPANESE ADVANCE, EARLY 1938
Yangtze River
Hankow
Ichang
LAKE POYANG
TUNGTING LAKE
Changteh
nkiang
Changsha
Nanchang
MAY 1938
Ling-ling
Hengyang
Sui-chuan
yun
Kweilin
Nan-hsiung
Peh River
Liuchow
Amoy
Si River
Swatow
Nanning
Canton
PE
Hong Kong

assemble them from British authorities in Burma (which may have been Pawley's influence towards a competitor).

When Hankow was threatened, William Pawley moved his operation to the Greenlaw factory in Hengyang, hoping it was out of bombing range at least temporarily. Pawley managed to assemble 30 Curtiss-Wright Hawk 75M fighters that had been ordered late in 1937 and had arrived through Canton before the Japanese seized that port. These new high-performance fighters went to operational squadrons which were consolidated at Chennault's new combat base in Chihkiang when that became available late in 1938. However, so many of these precious fighters were lost to accidents there that these squadrons were eventually sent to Kunming for further training.

Pawley decided to build a new factory as far from the Japanese as he could get. In November 1938, he borrowed one of Chennault's instructors, Billy MacDonald, to fly him further west along the Burma Road. They scouted a site for an airfield at Loiwing, on the Burma Road close to the Burma border. Pawley built a lavish new factory there. Some of Greenlaw's trainers made it to Loiwing for assembly before the British closed the Burma Road for three months in July 1940, but most remained in crates on Rangoon docks. Chennault had to continue training with whatever equipment remained, and student accidents steadily reduced that.

Warning network

Adequate warning was essential to intercept enemy bombers. In the fighting around the Yangtze valley, Chennault's warning net could use telephones for communication. But there was no telephone network in the mountains of southwest China. In the summer of 1940, he hired John Williams to set up a radio network in addition. This network would play a pivotal role in the battles to come.

Training pilots

The first classes of students at Kunming were difficult for the American instructors. The students started training in the Italian schools, and many were not skilled (even though they thought they were). The same problem (ineffective training leading to accidents) occurred in the operational squadrons being equipped with

Figure 1–2. Aftermath of Japanese raid at Liuchow, 1938. Student looks at the damage to a trainer. Many of the trainers were destroyed, forcing relocation. (Higgs)

Figure 1–3. Wrecked Fleet trainer at Liuchow, 1938. Liuchow was a good training location, with an existing airfield and hangar. It had previously served a local warlord's air force. Located near a rail line, essential supplies like aviation gas were available. Its loss was a blow to Chennault's training program for the Chinese Air Force. (Higgs)

the new high-performance Hawk 75 fighters at the new Chihkiang base. These squadrons were ordered to Kunming early in 1939 for further training. When the first group of 11 commissioned pilots flying their Hawk 75 fighters arrived in Kunming on January 27, 1939, three crashed upon landing!

Chennault's attempt to institute check rides with American instructors led to a mutiny in February 1939. The Chinese head of the training, Gen. Chow only settled this in Chennault's favor in March. The chaos of moving the basic and primary students from Liuchow further slowed progress. As happened each year, monsoon weather severely limited flying days during the summer.

Return to the US, 1939

Chennault got permission to return to the US in October 1939. During this time, he explored returning to military service there but the Army declined. No one in the Army Air Forces believed his reports of Japanese fighter performance. He visited the Curtiss-Wright factory in Buffalo, which was then ramping up production of the P-40. With more orders than the factory could fill, he had no chance of getting the latest US fighter for China.

Chennault spent Christmas with his wife and family in Waterproof, Louisiana, having been away for two and half years. He came back to China in February of 1940, discouraged and frustrated, but determined to resume his training duties. Something had to change for him to be of more help to China.

2

China Soldiers On

The topography of Asia greatly influenced the course of the Japanese invasion of China. In the north, Japanese armies from Manchuria fanned out and advanced south. Three north-south railroads there allowed easy supply, and separate army invasions followed their path. The flat North China plain provided ideal terrain for Japanese Army tanks, a weapon the Chinese could not counter. Only if the Japanese armies advanced too rashly could a Chinese army surround and beat them, as at Taierzhuang in March 1938.

Still, by June 1938, the Japanese army had crossed the Yellow River and were moving west from Suchow along the railroad on the south bank of the Yellow River. With no way to stop them from taking Chengchow and continuing south into Hankow, Chiang had the dikes on the Yellow River breached in June 1938. This unleashed a massive flood, with the Yellow River changing course to reach the ocean 200 miles south of its former mouth. Hundreds of thousands of Chinese villagers died as a result, but the Japanese advance from the north was stopped temporarily.

In central China, a combined Army-Navy invasion headed west from the port of Shanghai up the navigable Yangtze River. As they approached Hankow in the fall, another Japanese invasion force captured the port of Canton in October 1938. Losing this port closed

MAP 2–1. China's Land

Topographic features as China's defense

Manchurian Plain. Occupied since 1931, this region had coal and iron mines with much industry, and was the home base of Kwangtung Army.

North China Plain. Suited to growing wheat, this region also held major cities and industries. Flat dry ground allowed Japanese tanks and troops to advance rapidly.

Yellow River Flood. Japanese advance from Suchow toward Hankow was stopped in June 1938 by a major flood. A Yellow River dike was breached deliberately by direct order of Chiang Kai-shek. The entire river changed course and the flat plain allowed the flood to spread over a vast area, preventing army movements.

Lower Yangtze Valley. Most of China's major cities and industry were located along the Yangtze River due to its ease of navigation. The Japanese Navy used heavy artillery to squash resistance as their armies advanced up the valley in 1938. Big ships like cruisers and gun boats could lob shells up to 10 miles inland. Loss of Hankow in October of 1938 left the entire lower river, Nanking and Shanghai under Japanese control.

Szechwan Basin. This remote mountain fortress, circled by defensible mountain ranges, allowed the Chinese government and industry to retreat from the Lower Yangtze Valley to Chungking in October 1938.

Hunan Rice Bowl. Distracted by war with Russia over a disputed border in Manchuria, Japan held at Hankow, expecting China to surrender quickly. Frustrated after a year, in the fall 1939 Japanese troops moved south into Hunan. Here Japanese tanks encountered flooded rice paddies covering the valley floors. Tanks could advance only along roads, where they were easy targets for Chinese mortars from the surrounding hills. The Chinese defended Changsha for three years, stopping any advance to the interior.

Yunnan Highlands. This mountainous region, with rice paddies on the smaller valley floors, provided good defensive terrain for Chinese troops, but also made construction and use of the Burma Road difficult.

Southern Ranges. These lower mountains were still good defensive terrain for Chinese troops. Japanese movement north from Canton was limited by these mountains.

Canton. This major port city was the southern terminus of a north-south railroad running from Canton through Hankow to Peking. The Japanese seized control of the lines near Canton and Hankow in October 1938. However, the bulk of the rail lines between those cities remained under Chinese control, helping to distribute supplies within China.

MAP 2–2. RETREAT TO HANKOW, EARLY 1938

Japanese-occupied zones. Japan had long occupied Korea and Manchuria (Manchuko). In 1937, the Japanese army easily took the plains of northern China. Their navy used the Yangtze River to take Shanghai and Nanking late in 1937.

Japanese Armies Advance. The Chinese army held Suchow against a Japanese army advancing from the north. A second Japanese army advanced from the south and forced the Chinese to withdraw. By May, combined Japanese armies advanced west along the rail line toward Chengchow, threatening the temporary capital at Hankow.

Yellow River Flood. To block the army's advance south and secure the Russian supply route, in June, the Chinese breached the south dike of the Yellow River near Chengchow, unleashing a massive flood. While blocking Japanese advance, the flood also caused massive Chinese casualties and later, famine.

Soviet supply route. Trucks and camel caravans left from the Trans-Siberian Railway and covered almost 2000 miles getting to Hankow.

Hong Kong and Canton supply route. Japanese blockade of Canton by sea did not stop entry of supplies via British colony of Hong Kong.

Burma motor road. A truck road was under construction over mountainous terrain from Lashio, Burma, to Kunming, China.

Haiphong to Nanning. Supplies landed in the neutral port of Haiphong, French Indochina, were trucked to Nanning.

Hanoi to Kunming. The narrow gauge railroad carried military supplies to Kunming and Chennault's training bases.

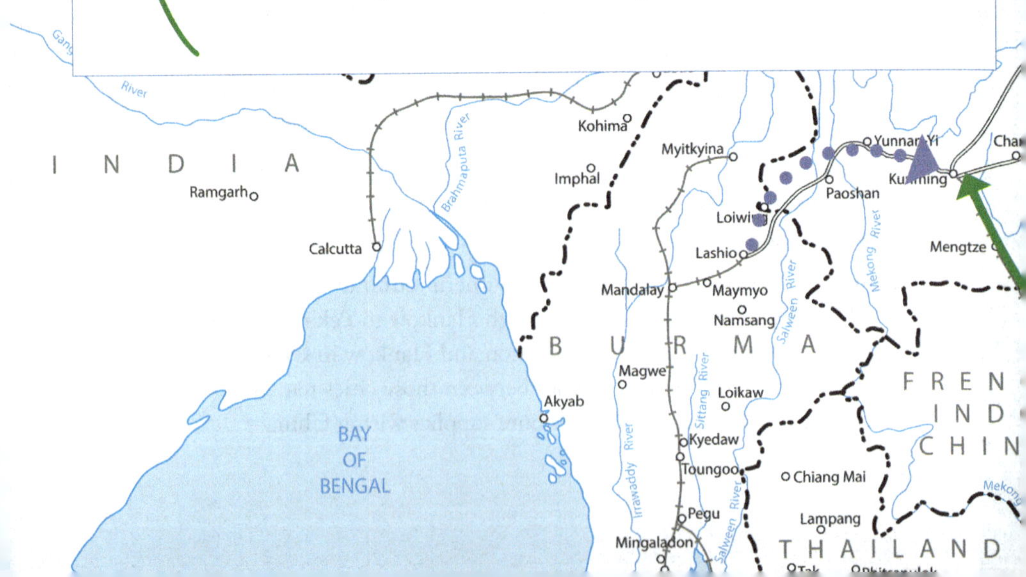

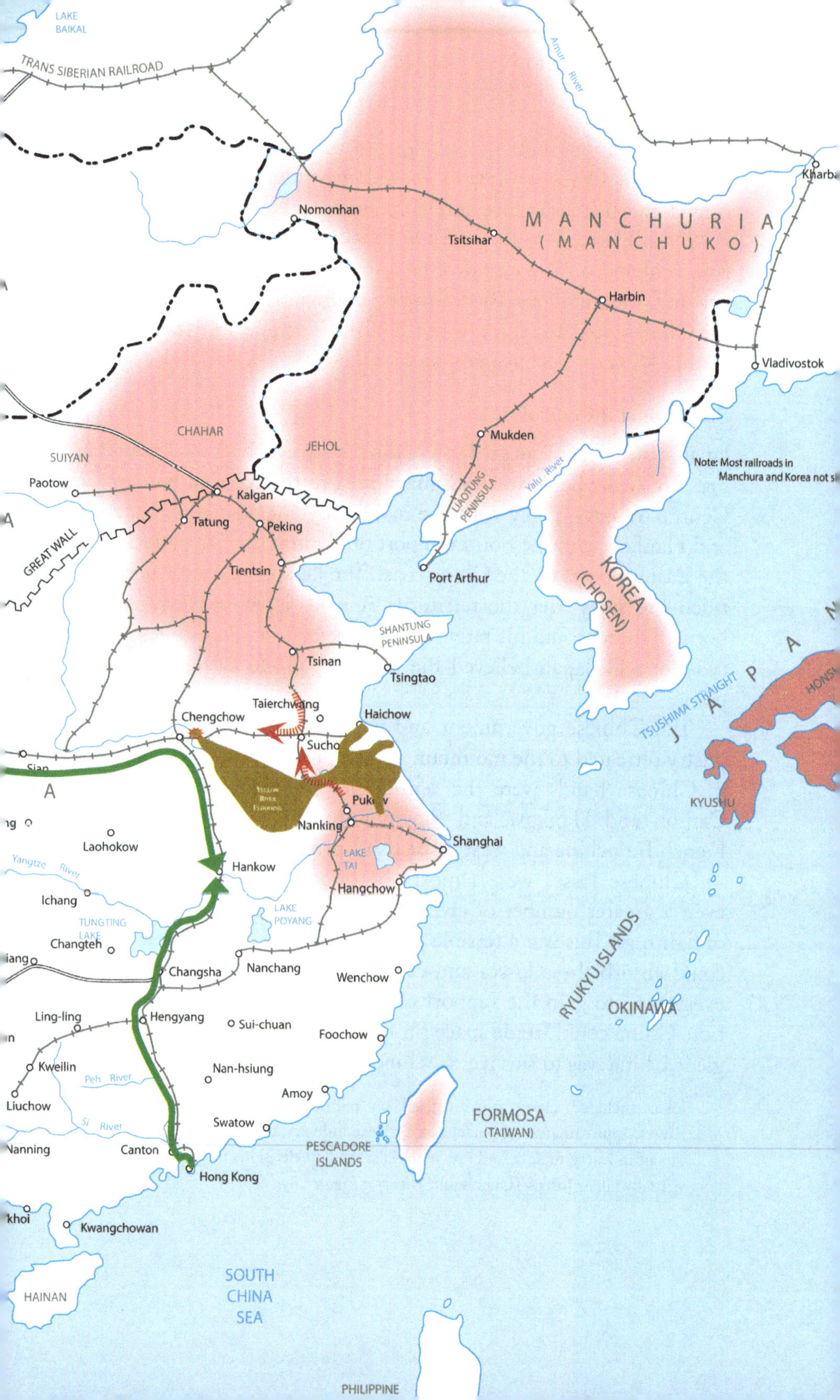

LAKE BAIKAL
TRANS SIBERIAN RAILROAD
Amur River
Kharba
Nomonhan
Tsitsihar
MANCHURIA
(MANCHUKO)
Harbin
Vladivostok
Mukden
CHAHAR
JEHOL
SUIYAN
Paotow
Kalgan
Tatung
Peking
GREAT WALL
Tientsin
LIAOTUNG PENINSULA
Yalu River
Note: Most railroads in Manchura and Korea not s
Port Arthur
KOREA (CHOSEN)
SHANTUNG PENINSULA
Tsinan
Tsingtao
Taierchwang
Chengchow
Haichow
Sucho
TSUSHIMA STRAIGHT
JAPAN
HONS
KYUSHU
Sian
Yellow River Flooding
Puk
Nanking
Shanghai
Laohokow
Yangtze River
Hankow
LAKE TAI
Hangchow
Ichang
TUNGTING LAKE
LAKE POYANG
Changteh
Changsha
Nanchang
Wenchow
RYUKYU ISLANDS
OKINAWA
Ling-ling
Hengyang
Sui-chuan
Foochow
Kweilin
Peh River
Nan-hsiung
Liuchow
Amoy
Si River
Swatow
FORMOSA (TAIWAN)
Nanning
Canton
PESCADORE ISLANDS
Hong Kong
Kwangchowan
HAINAN
SOUTH CHINA SEA
PHILIPPINE

the major supply route to Hankow, so the Chinese retreated from Hankow in October 1938, to the relative safety of the mountain-ringed Szechuan province and its major city Chungking.

From Hankow to the south lies a vast wetland used for growing rice, centered on the city of Changsha. Further south, the country becomes mountainous. By destroying roads and railroads in this area, the Chinese negated the advantage of the Japanese army tanks. Thus further Japanese advance was more difficult.

Situation at the end of 1938

Japan's initial tactical goals in China were reached by December 1938. Its Army held North China almost to the Russian border, the Central Yangtze valley with the major cities of Shanghai, Nanking and Hankow, and the Southern port of Canton. Its Navy blockaded the Eastern ports on the China coast. The Japanese Army Air Forces ruled the skies over the territories it held. Japanese Navy carrier-borne fighters could hit targets within several hundred miles of the coast at will. Japan believed the Chinese had to surrender at this point.

The Chinese government and what could be moved of its industry retreated to the mountain-ringed province of Szechwan. Still in Chinese hands were the central rice-producing areas between Canton and Hankow, and the mountainous provinces south to French Indochina and west to British Burma.

Chinese losses were almost 900,000 military casualties, with even a greater number of civilian deaths (all within a few months of fighting). This was a terrible price to pay, but Chiang knew China could absorb these losses and continue fighting. He was gambling everything to gain the support of western democracies or intervention. China could "trade space for time." He must keep up the struggle if China was to survive, as Chinese.

> losses included: eleven provinces, all key railroads, the Yangtze and Pearl Rivers, approximately 95 percent of Chinese industry, the capital and all centers of industry, finance and raw materials, the best divisions in the army and virtually all air forces. (Chennault, *Way of a Fighter*, 85)

The only advantages Chiang had in his fight against the Japanese were China's enormous population, the geographical size of the country, and the topographical barrier of the mountains that ringed Szechuan. All previous invaders succumbed to the vastness of China. Chiang felt with determined resistance the same fate would befall the Japanese.

> Although China's friends made extraordinary efforts, American isolationism remained stronger than sympathy. Polls showed only 2 percent of the public pro-Japanese against 74 percent pro-Chinese but these sentiments did not include a desire for involvement. (Tuchman, *Stilwell and the American Experience*, 189)

Chiang stated China would never surrender, regardless of Japanese actions. Considering the position China found itself in at the start of the war (1937), that was a bold statement. From this moment forward, Chiang vowed to fight until death. Chennault marveled at the courage and determination of the Chinese population, who were now united as never before to expel the Japanese invaders. Chiang now led a great nation.

However, continued fighting required outside supplies. Losing Hankow severed the rail connection with the long overland route from Russia. Russia continued to ferry pilots and aircraft into Szechuan, but the mountainous terrain and the lengthy distance limited ground transport of supplies such as ammunition and aviation fuel. Losing Canton severed the rail connection with the ports of Canton and Hong Kong. The neutral port of Haiphong in the French colony of Indochina became a significant source of supply. The port was connected to Kunming via a narrow-gauge railway, and to the interior of China by roads leading through the city of Nanning. The Burma Road brought supplies and aviation fuel from the port of Rangoon in the British colony of Burma.

Japanese long-term planning

For Japanese war planners, continuing their expedition into China's interior was an unexpected development. This was no longer a limited incident with a quick end in sight. Disagreement of how and where to continue the war caused ongoing friction in Japan. Debate

MAP 2–3. Retreat to Chungking, late 1938

Japanese-occupied zones. Japan held onto occupied territory and expanded its control south and west.

Japanese Armies Advance. The Yellow River flood bought six months for Chiang to move the capital from Hankow farther inland to Chungking. Heavy industries were also moved to Chungking, including entire factories. The flood waters receded but left the land unusable for growing crops. The southern ports of Amoy and Canton were captured by October, and the Japanese occupied Hankow.

Closed Supply Routes. These losses closed off two critical supply routes for the Chinese, the Soviet Union-Hankow route and the Canton–Hankow route. Hong Kong, while still British, was now isolated from the Chinese mainland. Hong Kong's airport still supported flights into China on private airlines such as CNAC and Eurasia.

East Coast Ports. Despite Japanese blockades, the east coast ports of Wenchow, Foochow, Swatow and Amoy still smuggled in cargo and transferred it by truck to interior rail lines outside of Japanese-controlled areas.

Burma motor road. Completed late in 1938, the Burma Road covered 700 miles of mountainous terrain to complete a supply route from Rangoon to Kunming. Chennault's forces, now based in Kunming, depended on these supplies to keep its planes and pilots flying.

Haiphong-Nanning and Hanoi-Kunming. The neutral port of Haiphong, French Indochina, funneled supplies to China via truck to Nanning and a narrow gauge railroad to Kunming. Chennault depended on this port for supplies as well.

Retreat to Chungking. People, industry and government moved upriver to Chungking in a mass exodus.

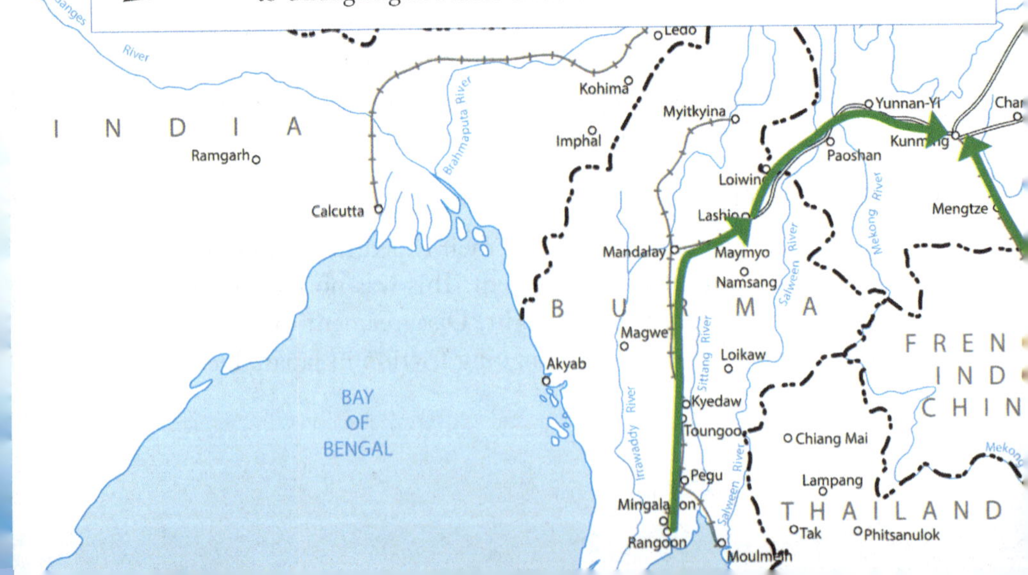

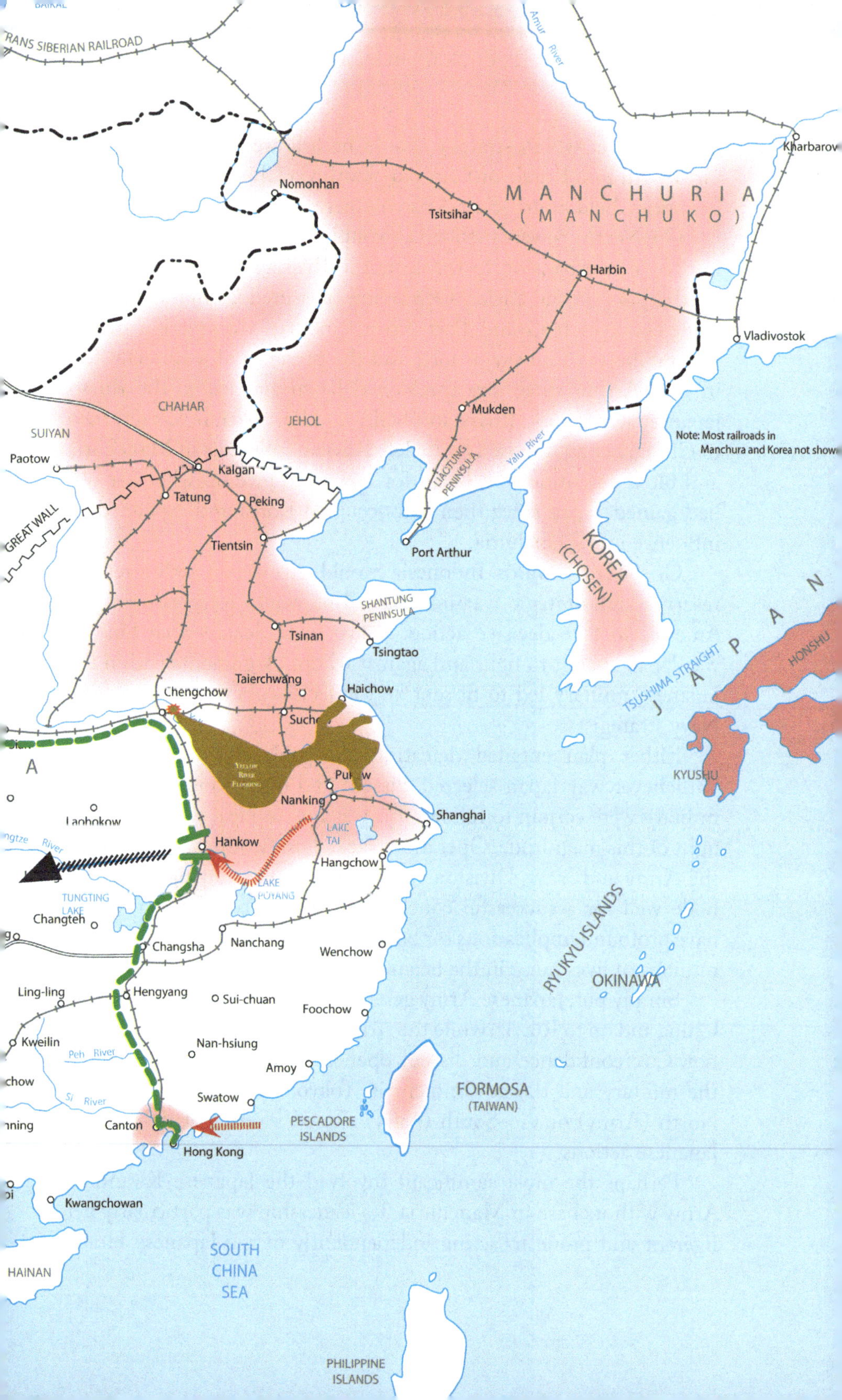

TRANS SIBERIAN RAILROAD
Amur River
Kharbarov
Nomonhan
MANCHURIA
(MANCHUKO)
Tsitsihar
Harbin
Vladivostok
CHAHAR
JEHOL
SUIYAN
Mukden
Paotow
Kalgan
Tatung
Peking
GREAT WALL
Tientsin
LIAOTUNG PENINSULA
Yalu River
Port Arthur
Note: Most railroads in Manchura and Korea not show
KOREA (CHOSEN)
SHANTUNG PENINSULA
Tsinan
Tsingtao
JAPAN
HONSHU
TSUSHIMA STRAIGHT
Taierchwang
Haichow
Chengchow
Yellow River Flooding
KYUSHU
Nanking
Shanghai
Laohokow
Hankow
LAKE TAI
Hangchow
LAKE POYANG
TUNGTING LAKE
Changteh
Changsha
Nanchang
Wenchow
RYUKYU ISLANDS
OKINAWA
Ling-ling
Hengyang
Sui-chuan
Foochow
Kweilin
Peh River
Nan-hsiung
Amoy
Si River
Swatow
FORMOSA
(TAIWAN)
Canton
PESCADORE ISLANDS
Hong Kong
Kwangchowan
HAINAN
SOUTH CHINA SEA
PHILIPPINE ISLANDS

followed over which strategy should be followed next: either Plan Go-North into Russia or Plan Go-South via China into the various European-held South-East Asia colonies.

Go-North towards Siberia would put an end to continuing Russian expansion to the east. The Russian Tsars had long desired a port on the Pacific Ocean and pressured China to concede Valdivostok in 1860 and Port Arthur in 1897. Completion of the Trans-Siberian Railway in 1904 and the presence of a Russian fleet in Port Arthur since 1872 threatened Japan and led to the Russo-Japanese war in 1905. Russia lost badly and Port Arthur was returned to China, but an imposed treaty left Russian influence in Manchuria and bitter resentment in Japan for having to give up what her war had gained. Japan since then had occupied Korea and expanded its influence into Manchuria.

Go-South towards Indonesia would give Japan unlimited oil reserves and strategic war materials. This would force Britain and America to take decisive action. Tokyo must decide what kind of war Japan sought to fight and against whom. Arguments within the Japanese military led to dissent and indecision between Army and Navy strategists.

Either plan entailed defeating a capable and strong enemy. Whichever way Japan selected, dissent and disharmony within the military were certain to follow. This was not an ideal way to begin a fight of this magnitude. Operations lacked unity of purpose between the Army and Navy. Division at the start of any campaign does not bode well for a successful conclusion. The choice of policy would have profound implications for both Japan and China. Millions upon millions of lives hung in the balance.

Simply put, Japanese Army generals favored a ground war across China and on to Russia, while the Admirals of the Japanese Navy did not. Overconfidence and dissent opened a substantial rift between the military and the government in Tokyo. Division in policy Go-North (Army) or Go-South (Navy) caused a series of unintended Japanese actions.

Perhaps the most significant involved the Japanese Kwantung Army with its base in Manchuria. Its leadership was particularly belligerent and prone to acting independently of the Japanese Home

Government. In 1931, this army had unilaterally seized Manchuria from the Chinese by force. In the summer of 1938, it clashed with Russian troops in a border dispute at Lake Khasan (south of Vladivostok). The Kwantung Army's defeat there only left it wanting to try again.

Japanese results in 1939

Early in 1939, the Japanese Navy invaded and secured the main ports on the Chinese east coast when the coastal Naval blockade proved too porous. They also seized Hainan Island (near the southern coast) in February 1939, and started to construct a significant air base there. From there, long-range bombers were able to raid Kunming in April and May, but there were more important targets.

Most Japanese long-range bombers then moved to air bases in Hankow, to attack the Chinese capital at Chungking and force a surrender. Chungking was beyond the range of the Japanese fighters so the bombers flew unescorted. The remnants of the Chinese Air Force inflicted enough damage that the Japanese shifted to night bombing when the moonlight was favorable. Chennault visited Chungking during the air raids. The onslaught continued throughout the sum-

FIGURE 2–1. CHUNGKING IN RUINS. Despite damage after repeated bombing raids, the Chinese in Chungking carried on. (WW2DB, 13376)

MAP 2–4. Stalemate in China, 1939

Japanese-occupied zones, 1939. The Japanese Kwangtung army waged a war with Russia from April to August 1939, but ultimately lost. Chinese armies successfully repelled a Japanese attempt to move south through Changsha. Southern and western China remained under Chinese control. Chennault was free to complete airfields, update training bases and train pilots.

Hainan Island Invasion. The Japanese Navy invaded Hainan in February 1939. They built an airbase and sent bombers to Chennault's base at Mengtze.

Changsha Battle. Chinese forces stopped a Japanese army advance on Changsha in September 1939. This preserved China's eastern airbases.

Nanning Invasion. The Japanese army and navy invaded Pakhoi and captured Nanning, closing the supply route from Haiphong by November.

East Coast Ports. By 1939, most Chinese seaports were held by the Japanese Navy and Army. Only Wenchow remained open, but was blockaded.

French Indochina to China supply route. One of the last open supply routes, Chennault's operation depended on the rail line from Haiphong to Kunming. French Indochina was neutral, so its port at Haiphong could not be blocked.

Rangoon to Kunming supply route. Providing aviation fuel, training aircraft and other goods, the Rangoon–Burma Road–Kunming route was crucial to Chennault in Kunming, and China.

Bombing Chungking. From May to August 1939, the Japanese Army air force bombed Chungking, flying from Hankow. Although the bombers were unescorted, the older Russian defensive fighters could barely match the speed of the modern Japanese bombers and were ineffective. Still, most raids were made on moonlight nights. Thousands of Chinese were killed.

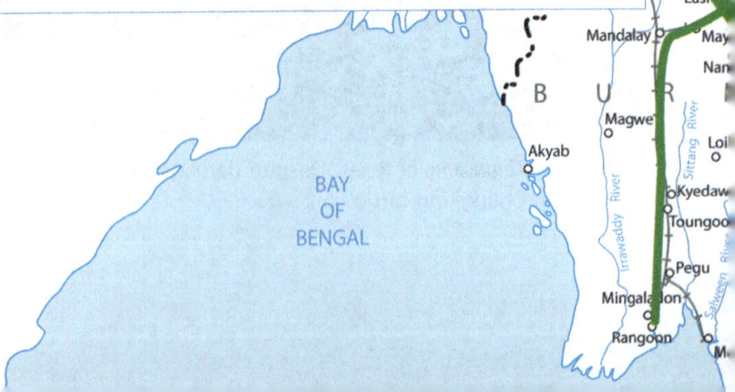

Nomonhan Battle. From April to August 1939, the Japanese Kwangtung army fought a major engagement with Soviet forces, ending in a Japanese defeat.
Irkutsk
TRANS SIBERIAN RAILROAD
Amur River
Nomonhan
Tsitsihar
M A N
(M A N
O U
O N
GOBI DESERT
ROUTE
CHAHAR
JEHOL
Mukden
SUIYAN
I N N E R
O N G O L I A
Paotow
Kalgan
Tatung
Peking
GREAT WALL
Tientsin
LIAOTUNG PENINSULA
Yalu River
Port Arthur
SHANTUNG PENINSULA
Tsinan
Tsingtao
Lanchow
Taierchwang
Haichow
Chengchow
Suchow
Yellow River
Sian
C H I N A
Pukow
Nanking
Shanghai
Ankang
Laohokow
LAKE TAI
Yangtze River
Hankow
Chengtu
Hangchow
Ichang
Chungking
LAKE POYANG
TUNGTING LAKE
Changteh
Chihkiang
Changsha
Nanchang
Wenchow
RYUKYU IS
Kweiyang
Ling-ling
Hengyang
Sui-chuan
Chanyi
Tuyun
Foochow
Kweilin
Peh River
Hungshui River
Nan-hsiung
Liuchow
Amoy
Si River
Swatow
FORMOSA (TAIWAN)
Mengtze
Nanning
Canton
PESCADORE ISLANDS
Hong Kong
Pakhoi
Hanoi
Kwangchowan
Haiphong
R E N C H
I N D O
C H I N A
HAINAN
SOUTH CHINA SEA
Mekong River
A N D
PHILIPPINE ISLANDS

mer. Here was the Douhet theory of airpower (the bombers will always get through) in action, and the results were devastating.

Meanwhile, in May 1939, the Kwangtung Army provoked a much larger battle with Russia at Nomonhan. This battle lasted the entire summer and diverted considerable Japanese army resources in men, tanks, artillery and aircraft, which otherwise could have fought in China. The Soviet Union committed even more resources to this battle, and in August General Zukhov's major counter-offensive annihilated the Japanese armies engaged there. The strength the Soviet Union displayed in inflicting this defeat and the signing of the Nazi-Soviet Non-Aggression Pact (also in August) played a significant role in Japan's eventual decision towards a Go-South strategy. In September, Hitler invaded Poland and World War II started in Europe.

While the battle raged at Nomonhan, the remaining Japanese Army forces in central China attempted to move south from the Yangtze valley in 1939. They seized the Chinese main airbase at Nanchang, but their forces did not defeat the Chinese armies at Changsha. This left south-central China (the major rice-producing area) under Chinese control and allowed internal movement of supplies being smuggled in from Chinese ports and from Haiphong in Indochina. The railroad from Hanoi to Kunming became particularly important, with tonnage tripling from 1936 to 1939.

Stopping the flow of supplies into China through Haiphong became a major Japanese focus in 1939. Diplomatic pressure on French authorities banned the transport of aviation fuel to Chennault in Kunming in June 1939. However, the French effort to restrict arms was half-hearted and supplies continued to flow surreptitiously. In response, the Japanese Army and Navy invaded the Chinese port of Pakhoi and advanced to capture the city of Nanning in November 1939. This blocked overland entry of supplies from Indochina into China. The Haiphong to Kunming railway and the Burma Road became China's remaining supply routes.

Japanese results in 1940

The Chinese launched a Winter Offensive throughout China late in 1939. Results were mixed, but Japanese forces stayed dispersed

Figure 2–2. Bombing Chungking. Japanese heavy bomber Ki-21, called a Sally by the allies, over Chungking. Bombs exploding show as white puffs. (ww2DB, 23816)

throughout China to counter this effort. In the south, Chinese forces dislodged the Japanese holding the Kunlun Pass near Nanning in January 1940. The Japanese army continued to hold Pakhoi.

In May, the bombing campaign against Chungking resumed with the onset of favorable weather. Japanese bombers included improved versions that were faster and could fly higher carrying a larger bomb load. The Chinese Air Force was hard-pressed to stop them.

Also in May, the Japanese began an advance up the Yangtze River toward Ichang, at the foot of the mountains surrounding Szechuan. Once they had captured the city and built an air base, their fighters could escort their bombers all the way to Chungking. In September, the long-range A6M Zero arrived. By December, this more capable Japanese fighter had eliminated any effective Chinese fighter defense.

In Europe, Germany invaded France which capitulated in June 1940. This left French Indochina without support, so the Japanese increased diplomatic pressure and threatened invasion to shut off the flow of supplies. Late in June, French officials closed the Haiphong-Kunming railway to military supplies, and even allowed Japanese ob-

OUTER MONGOLIA
CARAVAN AND MOTOR ROUTE
ROUTE
GOBI DESERT
INNER MONGOLIA
CHAHAR
SUIYAN
Paotow
Kalgan
Tatung
Tientsin
GREAT WALL
Ansi
Lanchow
Chengchow
Taier
Sian
Yellow River
CHINA
Ankang
Laohokow
Hankow
Yangtze River
Chengtu
Ichang
Chungking
TUNGTING LAKE
Changteh
Chihkiang
Changsha
Nanc
T
Sadiya
Ledo
Kohima
Imphal
Myitkyina
Yangtze River
Yunnan-Yi
Chanyi
Kweiyang
Tuyun
Ling-ling
Hengyang
Sui-chu
Kunming
Paoshan
Hungshui River
Kweilin
Peh River
Nan-hsiung
Loiwing
Lashio
Mekong River
Mengtze
Liuchow
Si River
Swatow
Mandalay
Maymyo
Namsang
Salween River
Nanning
Canton
Hong Kong
BURMA
Pakhoi
Hanoi
Haiphong
Kwangchowan
Magwe
Loikaw
FRENCH INDO CHINA
Akyab
Sittang River
Kyedaw
Toungoo
HAINAN
SOUTH CHINA SEA
Irrawaddy River
Chiang Mai
Pegu
Salween River
Mekong River
Lampang
Mingaladon
THAILAND
Rangoon
Tak
Phitsanulok
Moulmein
Tavoy
Bangkok
SPRATLY ISLAND 200 miles

servers in to verify that nothing was getting through. In September, Japanese landed troops anyway and set up air bases in Indochina, some within range of Kunming.

Similarly, with Britain hard-pressed in Europe, officials in Burma were wary of offending Japan. They yielded to Japanese pressure and closed the Burma Road to military supplies for three months starting on July 18, 1940, but reopened it in October after the monsoon ended. During the rainy season from June to September, road travel in Burma and up the Burma Road was difficult so this closure had less impact.

MAP 2–5. STRANGLEHOLD ON CHINA, 1940

Japanese-occupied zones, 1940. In May, the Japanese army moved west from Hankow to Ichang, establishing an airbase closer to Chungking. Invading French Indochina left the Burma Road as the sole open supply route.

French Indochina. Following the fall of France, the Japanese took over French Indochina in September, 1940. This prevented supplies entering through Haiphong, so they withdrew from Nanning. Japanese aircraft based in Hanoi could now reach Kunming. Chennault's office was destroyed in one of these bombing raids.

Rangoon to Kunming supply route. Providing aviation fuel and training aircraft, among other goods, the Rangoon–Burma Road–Kunming route was now the sole supply route to Chennault in Kunming, and to China.

Capture of Ichang. The Japanese advance to Ichang provided an airbase closer to Chungking.

Bombing of Chungking. Fighter protection using Zeros and upgraded bombers made the Japanese bombing of Chungking devastating. The Chinese Air Force was eliminated.

3

The Burma Road

Foreseeing the possibility of a Japanese blockade of its eastern ports, in early 1937 the central Chinese government began construction of a truck-capable supply route from Rangoon in Burma through the mountains to Kunming, China. This weaving track eventually became known as the Burma Road.

> In 1937, when the central government authorized the building of the "back door", China was still in control of most of the seaports and Shanghai had not been taken. The Canton-Hankow railway was still transporting goods and the French Indochina railway moved supplies to Kunming. Thus arteries within China were still fairly good. Yet it quickly became evident that a route outside China into Burma would become strategically necessary should Japan close the ports around the China Sea. The government decided to build the Burma Road from Kunming to Lashio in Burma where it could connect with the railway system at Rangoon. (Cornelius and Short, *Ding Hao*, 13)

This pathetic pathway later proved critical for China's survival. What they started with was anything but a road in the western sense, more like a mule trail hacked by the hands of thousands of Chinese peasants. Using conventional homemade tools fashioned out of rock, brass and locally smelted iron, the tedious task continued twenty-four hours a day. There were no cranes or heavy equipment of any kind except large hand-hewn rollers made of rocks, nine feet in diameter, and

Sadiya
Ledo
Kohima
Imphal
Myitkyina
Yangtze
Yunnan-Yi
Kunming
Paoshan
Loiwing
Lashio
Mengtze
Mandalay
Maymyo
Namsang
Salween River
Mekong River
BURMA
FRENCH INDOCHINA
Magwe
Loikaw
Sittang River
Irrawaddy River
Kyedaw
Toungoo
Chiang Mai
Pegu
Salween River
Lampang
Mingaladon
THAILAND
Rangoon
Moulmein
Tak
Phitsanulok
Tavoy
Bangkok

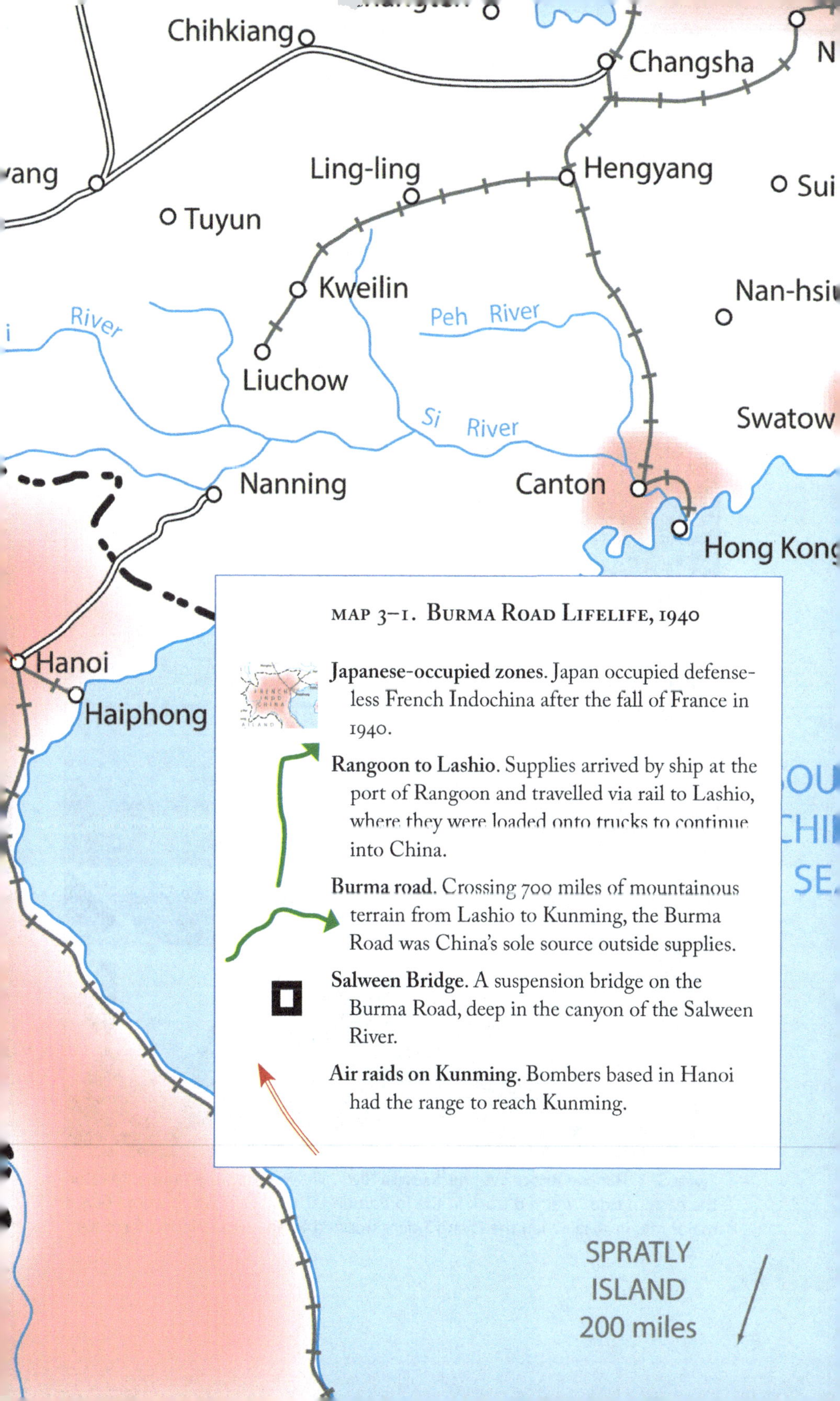

MAP 3–1. BURMA ROAD LIFELIFE, 1940

weighing 3–4 tons. From such humble beginnings, this "road" eventually became China's most vital lifeline to the outside world.

Determined by Chinese "engineers," the road track wove and turned. Local workers laid the right of way as they walked it. Measurements made by the eye determined angles, grades and turns. All problems were solved using district knowledge and experience. Two major rivers had to be crossed, the Mekong and the Salween.

Figure 3–1. Huitong Bridge over the Salween River. Viewed from the Burma Road on the eastern side. Wrecked truck in the foreground. This river crossing would play a major role in 1942, when the Flying Tigers stopped a Japanese invasion force here.

(US Army, cbi-theater.com, the Ledo Road)

The latter was particularly difficult. From its lowest point at the Salween River, the circuitous pathway climbed five thousand meandering feet along the edge of steep mountains.

In places there was a drop of a thousand feet or more straight down. Any western engineer or road builder would proclaim it an impossible task (the British did). Terrain was much too treacherous to attempt this kind of work. Heavy equipment, construction materials and supplies were not available to the Chinese. They must improvise everything (tools, materials, engineering expertise and design) necessary to keep construction moving.

British engineers working on the other end declared it could not be built without specialized grading equipment. This assessment did not even faze the Chinese. When difficulties presented themselves, they just picked up their tools and went back to work.

By the time the project finished, over 200,000 Chinese took part in its construction. It was these workers who "clawed out rock and dirt with their bare hands." They used crude spades and picks for digging.

Special crews of selected workers built two huge suspension bridges across the Salween and Mekong Rivers. Community experts engineered these bridges and built only with materials on hand. Everything needed was obtained from local resources.

Roadside smelters produced iron ore. Each link of the chain was hand cast and assembled in place, then loaded onto special barges. The swift-flowing river made the task even more daunting. Work crews assembled the links into chains. Hundreds of coolies pulling in opposite directions at the same time tightened the chains. Bridges were inspected on site and immediately deemed passable. Road construction continued without pause.

Thousands of Chinese men, women and children died while building this life line. Work never halted. Rain, snow, and blazing sun never hindered the effort. It was vital to open this roadway for supplies. Incredible feats of local construction defied established engineering techniques and allowed the lifeblood of China to flow uninterrupted for several more years.

The straight-line distance between Kunming and Lashio is 320 miles, but the new road would measure 717 miles, winding around such natural objects as

> mountains or obstacles created by man as well, like tiny, precious rice fields, national monuments, cemeteries, and temples to various gods, most of the road rested in China, only 117 miles was in Burma. The Chinese would build the road to the town of Wanting on the Burma-China border. The rest of the project was to be completed by the British government. (Cornelius and Short, *Ding Hao*, 14)

In no place was the roadway wider than twenty-three feet, the bare minimum for two trucks to pass one another at the same time. Handmade crushed rocks of differing sizes created the road bed. Coolies set graded stones in place by hand until they were seven to ten inches deep. The top layer was one inch stones held in place using a centuries-old slurry mix of mud and water, then set in place to dry. Stonemasons built retaining walls without the use of cement.

The Burma Road replicated the effort of building the Great Wall. It was just as important to the survival of the people of China. This supply line was a tribute to the courage and resourcefulness of the Chinese people.

During the Japanese blockade of Chinese ports, the Burma Road became China's last hope for overland resupply once the Soviets withdrew to fight Germany. Faced with ever mounting casualties and dwindling supplies, there seemed nothing China could do to save herself. If the Burma Road failed, China was doomed.

Japan was well aware of this fact. The Japanese penetrated steadily into China's interior, seeking to seal the route. Victory or defeat here would determine who would win the war in China. It was to protect this lifeline that led to the eventual formation of the American Volunteer Group (AVG), American pilots flying American planes to protect the Burma Road.

Figure 3–2. Truck convoy on the Burma Road. US Army trucks crossing a washed-out culvert in 1945. The road was barely wide enough for two trucks to pass, with sharp corners and frequent switchbacks. (US Army, cbi-theater.com, The Ledo Road)

Part Two

With the American Volunteer Group

4

Conception of the AVG

World events influenced the situation in China by mid-1940. Russia and Germany signed a nonaggression pact in August 1939. Then Germany formed the Tripartite Alliance between Germany, Italy and Japan in September 1940. With Japan now an ally of an ally, there was no further need for Russia to fight Japan in China.

The Russians began to withdraw their forces, being worried about Hitler's intentions, and this included the air force aid for China. With the Russian withdrawal in late 1939 and 1940, Japan no longer needed to maintain as significant an army in Manchuria to counter the Russians. China needed help from some other source. Chiang was desperate to protect the Burma Road and stop the bombing of Chungking.

In June 1940, Chiang sent Dr. Soong to Washington to lobby for increased aid for China, including military equipment. Only the United States could provide additional resources quickly, but Washington moved slowly. So in October 1940,

> ...Chiang summoned Chennault to his headquarters and instructed him to proceed to Washington to lobby for the sale of American airplanes to China for immediate relief from Japanese bombing runs. (Yu, *The Dragon's War*, 31)

During his career in the Army, Chennault had established a reputation as an "out of the box" thinker. The flip side of this from the view of Army brass was a reputation as an insubordinate, argumentative rebel. There was also another factor working against Chennault. Chennault had originally joined the Army via the Reserve Officer Training Corps program. He had applied, but had decided not to go to the Army Service Academy at West Point to become a regular officer.

Graduates of the Academy wear class rings as a sign of their status, whereas lower-status reserve officers do not. If a senior graduate of the academy felt a discussion had gone on long enough and wanted it finished agreeing to his position, he would knock his class ring on the table. All other Academy graduates were expected to hew to his position. Non-Academy officers resented this status and power difference. Academy graduates became known derogatorily as "ring knockers."

> Lauchlin Currie and T.V. Soong brought some notable visitors to my home... The other visitor Soong brought around to my home was known to me by reputation—and a somewhat checkered one at that. Considered incorrigibly insubordinate by his recent superiors, he was one of the most remarkable soldiers I was ever privileged to know: Claire Chennault. (Corcoran, *unpublished Biography)*

It seemed to Chennault as though the ring knocker faction would never let up trying to squelch his ideas.

Appeal to Roosevelt

By the fall of 1940, President Roosevelt was worried about the success of the axis powers Germany, Italy and Japan. So military support for China seemed like a better idea than it had previously. This time Chennault got a different reception in Washington.

> Through Missy LeHand [Roosevelt's gatekeeper] I'd received orders from the President to meet this singular man and seize him up. Was he a tactical genius? Or just completely out of step with his erstwhile colleagues in the Army Air Corps? (Corcoran, *unpublished Biography)*

And so it went up through various chains of command. Where there was resistance, the word came down from the top. Eventually the dust settled and Roosevelt got his way—military help for China that was not obviously tied to the US Government. The deal was this: 100 new P-40Bs fighters, called Tomahawks, from the Curtiss-Wright factory in Buffalo. Pilots and ground crew were allowed to resign from military service and join the Chinese Air Force. They would be generously paid through Pawley's CAMCO company and a new entity called China Defense Services (CDS).

> Money wasn't so scarce in Washington. CDS had cash enough for whatever costs it had to cover at any moment. The problem was to locate and acquire the goods we wanted to buy: weapons, fuel, ammunition, vehicles, spare parts, medical supplies, and, above all else, planes. Then pilots! (Corcoran, *unpublished Biography)*

Planes and personnel traveled by ships to Rangoon. Unlike Greenlaw's troubles getting permission to assemble trainers in Burma, Pawley's company CAMCO quickly secured a place to assemble the fighters at the main Rangoon airfield, Mingaladon. Pawley had ways of getting what he wanted. The British offered up an airfield called Kyedaw at a Burmese town called Toungoo, as a base for training. CDS hired several recruiters to visit military bases to sign up pilots and ground crew.

The tar-ball

Chinese soil was no longer black but soaked red from the blood of her people. Japan won each battle but was no closer to victory. What could the Japanese do to get this war finished? A tar-ball syndrome caused concern for Japanese field commanders. Now engaged, they could not get free. Just like Grant in the Wilderness Campaign, the Chinese army held on tightly.

Good defensive lines mean lots of offensive casualties. Chiang did his math. He told his generals not to give the Japanese any cheap victories. Draw them deeper into the interior. This was magnetic warfare [as Chiang called it] at its best. Chiang's strategy encouraged the Japanese Army to follow his retreating forces deeper, ever deeper into China. Japan seemed just short of their moment of victory. Tokyo thought one more determined push would finish the Chinese once and for all.

The Japanese were fighting in south Henan in January 1941. By now, they controlled all coastal areas, inland pockets and wherever large towns or cities were located throughout coastal China. Total victory must certainly come soon. Japan did not have adequate strength to control the vast interior. Short of manpower, it was all the Japanese Army could do to hold the ground they stood on. Fighting became siege warfare in reverse. The Japanese Army only ventured outside of their protected enclaves to conduct major field operations or to find food. Whenever the Japanese Army deployed too small a force away from their redoubts, Chinese guerillas and irregular forces swept down from the mountains and destroyed them. These were punishment raids, no quarter given.

In the northern interior, a young Mao Tse-tung perfected this type of raid into an art form. These were his missions of choice. Tactics were always the same; strike unexpectedly, fight hard, destroy the enemy. Take no prisoners, carry away all weapons and usable equipment, then taunt the enemy forward in pursuit. Mission accomplished, these irregulars returned to their mountain hideouts and safety.

Here was reality for the glorious Japanese Army. This was not warfare to dazzle the folks back home. Battle was a slow, painful, bloody affair. In China, the Japanese learned a hard lesson. German soldiers learned the same painful lesson and would come to the same conclusion in Russia in 1941–1942 at Moscow and Stalingrad.

War in China discouraged Japanese soldiers. Seeking sweeping, dynamic victory, determined Chinese resistance thwarted Japan. This infuriated Japan's field commanders. Slaughter in the interior continued unabated for the invaded. The bloody year of 1941 began with no substantial change in war fighting. The battles of Western Hopei and Shanggao did not bring the expected Chinese surrender. Japanese field commanders now questioned the validity of going any deeper into China.

Bitter war dragged on. Chinese forces resisted as best they could. Slowly, they retreated to the interior. Like Grant at Petersburg, Virginia, in 1865, Chiang's strategy was engagement without respite. China would trade space for time. The combatants lost many hundreds of thousands of soldiers. The Chinese continued to lose battle after battle. The sacrifice of her soldiers continued to mount. Chiang knew China could afford these losses. Could the Japanese Army afford them also?

No end in sight to this meat-grinder war Japan increased strength to over one million men. After all the effort and material expended in China, it was time to win this war once and for all. The Japanese General Staff knew China must fall soon for other scheduled operations to begin later in the year.

The imponderable mystery of China was unexpected: their unforeseen strength and unity. This caused concern for the Japanese High Command. Determined to resist at all costs, Chiang continued fighting. China would bear any loss, be it men, material or treasure.

How much could the Chinese people take? The more the Japanese punished, the more the Chinese endured. When would Japan's victory come? What did it take to make these obstinate Chinese accept defeat? It was time for new strategies, both for Japanese invaders and Chinese defenders. The success of these opposing strategies would determine which great empire triumphed and which perished from the face of the earth.

5

Birth of the AVG

Many excellent books have been written over the decades about the First American Volunteer Group (AVG). Some true, some not so much but all interesting. The pilots and ground crew were a true cross section of America and came from every corner of the continent. Some were young idealists, others middle-aged realists. A few were from patrician families and many were from humble immigrant backgrounds. Most were brave men, a few were cowards. Many were honest, a few were thieves. Some were running to something, others were running from something. Most were trying to prove themselves, while a few were seeking to hide. Some became heroes, while others sank into obscurity. Whatever they did from the moment they signed their CAMCO contracts, they became an indelible part of America's aviation history of the Second World War.

A few believed this was a career enhancer and might get them regular commissions. Others saw it as a chance to fly fighter aircraft in combat. Some misfits thought this was an easy way out of their military commitments. Most wanted to volunteer, but some were volunteered (gotten rid of) by their commanders.

All were changed forever by their commitment. Incredible courage and success made them heroes around the world. Stories about them, both real or enhanced, enshrined them among the annals of great airmen of the ages. Time passes quickly and lines between truth and fiction blur easily. Let's see what we can find.

To provide the reader with a true sense of who these men were, where they came from and what they accomplished, I have quoted as much as possible from their own records, diaries, letters, books and recollections. Personal documents formed the source material whenever possible. This gives us an accurate portrayal of the men, their motivation and their actions. Charley Bond recalled:

> I had my own reasons for volunteering...The late night phone call ...from my old roommate...Jebbo was aware of my intense desire to become a fighter pilot and to secure a regular commission for a career in the Air Corps... A retired pursuit pilot named Claire L. Chennault had obtained personal approval from the President to recruit pilots and mechanics from the U.S. armed forces for duty with an American volunteer group. (Bond, *Flying Tiger's Diary*, 19–20)

Tex Hill remembered it this way:

> President Roosevelt's secret executive order authorized military pilots and ground crew members to resign their positions and serve in the new American Volunteer Group. This prompted Chennault to send a team of picked recruiters to every major Army Air Corps base and Naval Air Station in the country. (Hill, *"Tex" Hill: Flying Tiger*, 69–70)

This fit with the intelligence they had seen. Senior naval personnel observed signs of growing Japanese maritime military activity. Many believed war with Japan was inevitable. That was all David "Tex" Hill and Ed Rector needed to know. It was time to go! Flying fighters couldn't be that different from flying dive-bombers, could it?

> China had been fighting the Japanese for the previous four years and whose military forces had been pushed back eight hundred miles into the interior, was in grave danger of collapse. (Howard, *Roar of the Tiger*, 57–58)

James H. Howard was the son of a missionary in China and was going back home to China. He had personal scores to settle.

Not all who volunteered for this experiment in democracy were patriots. Greg Boyington joined for the money. Divorced and in debt, Boyington was under orders from the Marine Corps to repay his creditors or face disciplinary action. Each month, he must account to H.Q. the amount of money paid on each of his many debts.

Out of fear of not being "chosen" for this high-paying assignment, Boyington neglected to tell the CAMCO recruiter about his financial problems. He did not consider it necessary to discuss this matter with his new employer. Nor did he tell him he was the life of a cocktail party, either. The AVG found out soon enough. Boyington was not the only one who liked to drink too much.

As an officer in the military, there are two things you must never do: accrue unpaid debts or appear in public in a drunken, disorderly state. Boyington did both frequently. This is a violation of military decorum and considered a poor reflection on the entire service.

It was unfortunate for Boyington, the Marine Corps and the American Volunteer Group that he could not control his spending or drinking. Boyington became a source of much contention and adverse publicity about the AVG. Some of his pranks could not be written off as a young officer having fun (especially in China).

Later he became famous as Pappy Boyington of the Black Sheep Squadron on Guadalcanal, a Japanese prisoner of war and a Medal of Honor winner. Alcohol problems tarnished his reputation. It was a shame Boyington had this stigma because he was the stuff from which legends are made. Fellow Guadalcanal Marine pilot and Medal of Honor winner Joe Foss said of Boyington,

> Pappy was the nicest guy in the world when he wasn't drunk; unfortunately he was always drunk. (Foss, *Conversation with the author*, 2001)

R.T. Smith learned of the AVG by accident.

> It was in a casual conversation at the bar in the Officer's Club….I was talking with one of the older instructors….asked him if he knew anything about an outfit being put together to go and help defend the Burma Road on behalf of China. (Smith, *Tale of the Tiger*, 17)

A few days later, they had an interview and some drinks with the AVG recruiter. Next thing they knew, both were off to San Francisco. God only knew what awaited them, but it was exciting.

> They [recruiters] carried letters authorizing freedom to talk with personnel, after which they offered one-year contracts to "manufacture, repair, and operate aircraft" for CAMCO. Monthly salaries ranged from $250 per month for mechanics to $600 for pilots and $750 for squadron commanders. In addition, traveling expenses and quarters would be furnished, and they would receive

> one month of paid leave. Although it was not stated in the contract, recruiters hinted that pilots could earn a $500 bonus for each confirmed enemy plane shot down. (Bond and Anderson, *Flying Tiger's Diary*, 18)

This was the way it was for most of the volunteers. The money was typically two to three times what they were making in military service.

Many recruiters did their jobs diligently, trying to find the best candidates possible. There were enough problems associated with this endeavor without worrying about personnel. Most recruiters tried to find qualified people–most recruiters, but not all. After hundreds of interviews, some men took men who were not the right sort. This was unfortunate and took great effort by Chennault to sort out once overseas.

> Chennault was feeling excitement after the long months of desk work. His dream was coming true. On July 8, 1941, he flew from Washington to San Francisco and met with Aldworth and Tom Gentry—who was to be the flight surgeon for the AVG—and inspected the SS *Jagersfontein* before she sailed. (Samson, *The Flying Tiger*, 86–87)

Although it was a long time coming and faced almost insurmountable odds, the job was finally done. Credit Chennault and his few supporters that the group came together so quickly. Score another miracle for the retired Army Captain.

The gang's all here

Chennault now had crews, airplanes and supplies. The British in Burma allowed them an auxiliary airfield called Kyedaw at the town of Toungoo for them to call home. All that remained was to get them to the right place at the right time. The only way to transport people halfway around the world in 1941 was by the proverbial slow boat to China. In this case, four slow boats. Transit for most of the AVG took at least fifty days.

By May, recruiters provided 128 pilots and fifty-eight ground crew, as well as two doctors, two nurses, one dentist and two technicians. By September, fifty-five administrative personnel were on board. Eventually, 289 people joined the AVG and made ready to board ships as they became available. All went on various types of cruises. The lucky ones traveled in first class luxury, the not so lucky traveled under more strenuous conditions.

The first group sailed on June 9, 1941, aboard a military transport vessel, the unlucky ones. Shipping out with them was a contingent of Regular Army troops bound for Manila. The casual status, ex-military personnel of the AVG were under officers of the Regular Army, with a physical fitness regimen, canned food and all. These men traveled under the guidance of Paul Frillman, a Lutheran minister who spoke fluent Chinese. Chennault first met him earlier in Hankow and suggested he would be a good chaplain for the AVG.

Frillman was already earning his keep. Acting as sponsor of a wild bunch of airmen onboard for the long trip to the Orient was a difficult task. He realized he couldn't exert too much authority over his charges when they began referring to him as "Holy Joe." The troopship took the group from San Francisco via Australia to Manila. Frillman served as ringmaster and liaison between their military hosts and noisy, irreverent AVG men. This contingent of thirty air-

Figure 5–1. Dutch Passenger Steamer *Jagersfontein.* The second group of AVG volunteers traveled in style. (Neale, 243)

men arrived in Rangoon on July 28,1941 aboard an ancient Dutch coastal freighter.

Next to sail from San Francisco departed on July 10 on board the Dutch passenger steamer SS *Jagersfontein*, the lucky ones. This contingent of 123 traveled in first class luxury, white linen tablecloths and wonderful meals. Members of Chennault's medical team were also on board. Army flight surgeon Dr. Tom Gentry oversaw two doctors, a dentist and two flight nurses. Like everyone else involved with the AVG, the medical team was made up of free spirits and adventurers. The ship's bar remained open all day, and the AVG took it over. The only complaint from the men was the rest of the passengers were religious missionaries on their way to preach the gospel in China. Many men attempted to convert their fellow lady travelers to AVG standards but had no luck.

On the voyage, there was constant banging on the piano. Missionaries sang religious hymns followed immediately by wild drinking songs from AVG men. Both groups were happy to see the voyage end. On board for this never-ending party, the pilots, medical personnel and administrative types grew close. Thus began the start of the 1st AVG identity as hard drinking, irreverent fighters. They redefined the shipboard term of "an open bar," the AVG called it the "never closed bar."

Pilot Erik Shilling recalled their departure

> As we sailed outbound, a group gathered on the ship's fantail to look at the Golden Gate Bridge as our ship passed under it. Entranced, all of us watched as it gradually faded from view. The Golden Gate was a symbol of freedom for many immigrants coming to the states. In contrast, it was our last tangible tie to our country that we loved, and I will cherish the memory of the sight as long as I live. Words were unspoken, but in our hearts, many knew that some among us would not live to see it again. Who? Only time would tell. (Shilling, *Destiny*, 63)

The group on board thought they were traveling under tight security, right? They were supposed to be on a top secret mission. An enterprising young dockside reporter wrote a very comprehensive story about the secretive band of men sailing overseas. The novice scribe elaborated their reason for going abroad, the name of their ship, route, and estimated arrival time in Rangoon. That was pretty good reporting! This breach of security forced Lauchlin Currie to suggest President Roosevelt should order some type of escort protection for the ship. It would be a terrible start to this venture if the AVG lost over one hundred pilots and mechanics to Japanese torpedoes.

The *Jagersfontein* rendezvoused off Hawaii with the cruisers USS *Salt Lake City* and *Northampton*. They safely escorted the ship to Australia. The Japanese did indeed know about this trip and announced over Radio Tokyo they planned to intercept and sink the vessel. Loose lips really could sink ships.

On August 15, 1941, the *Jagersfontein* reached the Irrawaddy River outlet and sailed up to Rangoon. They crept into port late that night amid torrential thunderstorms and pitch black skies. This setting somehow seemed appropriate.

Another Dutch passenger liner left San Francisco on July 24, 1941, bound for Rangoon, Burma. AVG men were on board for their fifty-day pleasure cruise to Asia. The SS *Bloemfontein* brought an addition-

FIGURE 5–2. *NORTHAMPTON* FROM *JAGERSFONTEIN*. Two US Navy cruisers accompanied the AVG volunteers across the Pacific to defend against threats from the Japanese navy. Not many passengers rated a military escort. (NEALE, 246)

al twenty-seven Chennault recruits for foreign service. These transshipped to a small, rusty island-hopping vessel, the Penang Trader, for the final leg of their journey. Like earlier recruits, after fifty plus days of hard drinking and sumptuous meals, they arrived hung over and out of shape. These were among the last of the not-ready-for-prime-time flyers to arrive. Chennault needed to whip them into shape in a hurry.

Sailing on September 24, the liner *Boschfontein* carried an additional twenty-six AVG members. Traveling under the guise of missionaries, they put the fear of God into the crew and other legitimate missionary passengers on board. Genuine missionaries wondered what kind of gospel these men were going to preach. Arriving in Rangoon on November 12, 1941, they were weeks behind schedule for training and simulated combat flying. These men didn't yet know how soon they would fly in real combat. Just three weeks!!

> Three hundred of us, sailors, marines, and Army Air Force, had received our discharges from the service and in contingents numbering from thirty up to a hundred had traveled from California to Rangoon during the last four months. The armed forces resented having to let us go, but President Roosevelt's order could not be countermanded. If a man volunteered and was accepted, he had to be given an honorable discharge and sent on his way. (Trumble, *Biography)*

What the boss thought

First impressions can mean a lot. Chennault was the master of creating a strong impression. Paul Frillman observed this characteristic from the deck of the Dutch freighter when it docked in Rangoon.

> I don't suppose anyone could have called Chennault "glamourous" to his face without being punched. But he was a vain man and obviously relished making an impression. Like MacArthur, he had immense natural magnetism on which he based his public figure. (Samson, *The Flying Tiger*, 89)

Physical appearance, exotic dress, and that stern, almost fearsome face presented exactly the impression Chennault wanted to convey. There was no doubt about who was in command when Chennault was present. Every word he spoke was measured and precise, meant to convey his exact message. No one in the group ever questioned his ability, knowledge, or authority. What Chennault said, he meant, and everyone knew it. Around the Colonel you kept your ears open and your mouth shut. Like his students at ACTS, the men of the AVG hung on every word Chennault spoke.

When the final contingent arrived in Rangoon, Chennault was not there to meet them. He was in Chungking, conferring with the Generalissimo and his Chinese air staff. Chennault took a scheduled CNAC flight to Rangoon and then flew back to the Kyedaw base at Toungoo in a twin engine Beechcraft.

> The ground crews and pilots had been sweating and stewing in the heat and humidity and were all in a foul mood. Toungoo, a shabby little Burmese town alongside a railroad track and bordering the Sittang River, was lined with corrugated tin-roofed shops of Chinese and Indian merchants. Except for a dirt road-which alternated between being mud and dust-down the center of town, Toungoo was surrounded by steamy jungles and teakwood plantations. (Samson, *The Flying Tiger*, 93)

The heat in the jungle was stifling and humid. Most men thought it was unlivable. For Chennault, a native Louisiana boy, the weather reminded him of pleasant summer days spent living in New Orleans. Most of the other men had not experienced that particular pleasure. They thought they were living in hell on Earth.

Toungoo

Base facilities, such as they existed, were good quality, solid teak barracks, no screens, cold showers, plenty of bugs and minimum comforts–in other words, a typical RAF base.

> The climate, of course, was something none of us had ever experienced before, and it took its toll on both health and morale. The constant heat and humidity were enervating, to say the least, and coupled with the abundance of rain, led to frequent summer colds and sinus problems. Malaria and dengue fever began to crop up among a few of the more unfortunate; we learned we were in the middle of one of the worst areas in the world for those two mosquito-borne diseases. Dysentery and diarrhea were a couple of other less charming illnesses that beset most of us with embarrassing frequency. (Smith, *Tale of the Tiger*, 74)

Burmese cooks served food purported to be edible, but that was a matter of contention with the men. Even now amongst AVG men, food is still the main complaint about their first days at Toungoo. Arriving during the wet season was bound to slow down training. It rained for several hours each day. Airfield, roads and town were one continuous, gooey mess, more like quicksand than soil. Be it ever so humble, there is no place like home, and this was it!

> The heat didn't ease up while we were in Burma. Inside the heavy mosquito nettings enclosing our cots, the sweat streamed off of us, soaking the mattresses. If you got drunk enough before you turned in, you'd sleep. Seldom a breeze and even a breeze couldn't penetrate the heavy lacing of mesh which kept the mosquitos and myriads of other insects from making a feast on your body. Eventually you'd doze off and reveille at dawn and drinking hot coffee outside the cook shack was a welcome change from a restless night. It was fairly cool then, by contrast of course and that became the friendly and relaxed time. (Trumble, *Biography)*

Figure 5-3. Street in Toungoo. Toungoo was the town closest to the Kyedaw airbase, so Americans visited there during free time. (Neale, 249)

Toungoo was the quintessential jungle town, complete with cobras, other venomous reptiles, spiders, and many biting insects. It was located north of Rangoon on the narrow-gauge railway serving northern Burma, Mandalay and the Burma road terminus at Lashio.

Chennault had hired Harvey Greenlaw as his executive officer. His wife, beautiful, feminine Olga Greenlaw, became the official AVG historian. In her book about the AVG, she describes her arrival:

> After a couple of days [in Rangoon] Harvey wired me to come along to Toungoo on the "Up Mail." I can brush Toungoo aside with a few well-chosen expletives. There was a main street, rumbling day and night with trucks bound for Lashio and the Burma Road; a bazaar with shops on one side and open market stalls on the other; perhaps forty Europeans all told—and all the bugs God created to fly through the air or crawl on the ground, floors, walls, ceilings, into your food, down your back, up your legs, and in your hair: beetles, lice, spiders, flies and fleas, moths, mosquitoes, centipedes, bed-bugs, ticks, and a lot more you never heard of. The place was an entomologist's paradise. (Greenlaw, *The Lady and the Tigers*, 30)

How could tough military men complain when a beautiful woman went through exactly the same experience and she was wearing lipstick and high heels?

Rebellion

Just when Chennault thought he was getting a handle on his preparations, Chiang called him to Chungking.

> He found himself trapped for two weeks in Chungking during the worst bombing raids he had experienced in his three years in China....he was trapped in a house for seventy-two hours with no water, electricity, or telephone service, surviving on cold rice, and eventually forced to take shelter in a cave. Japanese bombers were overhead daily from dawn to dusk. (Schultz, *The Maverick War*, p. 106)

Chennault returned from Chungking on the twenty-first and received resignations from five pilots and several ground crew. All wanted to go home, citing misrepresentation as their reason for departure.

> To forestall mutiny, Chennault sensibly accepted, signed the resignations, and sent the men packing. He then called for a meeting of all personnel. Reverting to his old Army Air Corps days, he became a hard-nosed commanding officer. (Samson, *The Flying Tiger*, 94)

Chennault went back to his Bible for a solution. He believed that idle hands were the devil's workshop. To preclude anymore misguided attempts to get out of their contracts, Chennault designed a strenuous program of classroom study with a vigorous physical fitness regime. Just as his father did with him after his mother's death, Chennault worked them almost to exhaustion. Now they were too tired to think of other things, like going home. Complaining he could take, but desertion was no longer an option.

> Staring down the disgruntled volunteers with his snapping black eyes, Chennault told them there would be calisthenics, a regular routine, and fines for misbehavior. Meals would be served at regular hours and the men would be fined for being late for anything. They were to shape up. There was a sullen silence as they filed out of the hall, but no one had any doubt who was in charge. (Samson, *The Flying Tiger*, 95)

All pilots would take sixty hours of training beginning immediately and everyone must attend ground school. These classes consisted of two courses of seventy-two hours each-beginning the following morning at 6 a.m.

Figure 5-4. P-40 at Kyedaw Hangar. Planes were assembled at Mingaladon, flown to Kyedaw and completed there. This plane awaits installation of guns in wings and nose. The hangar was open on both ends due to the summer heat in Burma. (Neale, 250)

6

The Newbies

The Aircraft

Many airmen coming to fly for Chennault never saw the P-40 or had any idea how it flew. In combat, only two things (except luck) matter, skill of the pilot and performance of the airplane. Most of the men only heard stories of the P-40, and many of these were not a positive reflection of performance in combat.

> The pilots had as little regard for Curtiss-Wright's P-40 at Toungoo as I did, but for different reasons... Before the pilots left the United States, the P-40 had acquired a reputation as a killer in the hands of a relatively inexperienced pilot. (Chennault, *Way of a Fighter*, 115)

Not completely true of the P-40, but it was the impression his pilots had. By nature, flyers are a superstitious lot and this is not the best way to approach combat. To overcome any prejudice, Chennault ordered increased flying in the airplane until the pilots became comfortable. His prescribed flying training program did that. By the time AVG pilots saw combat, they believed their P-40 was a good plane. It would bring them through combat and back home again.

Chennault's initial opponents were the Nakajima Ki-27 "Nate" and the Nakajima Ki-43 Hayabusa "Oscar." Both were good aircraft for the time and in many ways outclassed the Curtiss-Wright P-40.

FIGURE 6–1. TWO P-40S IN FLIGHT. Typical weather in Burma, 1941. (NEALE, 199)

The Ki-43 was also commonly misidentified as a Zero early in the war, as did Chennault in his diary and book. Later, the real Zero arrived on the scene at Chungking, and simply outclassed the Russian fighters. No Zeros were fielded in Burma, but Chennault's tactics worked on the Oscars and Nates as well.

> Chennault compared the P-40 to the Japanese Zero. Designed in 1939, produced by Mitsubishi, the Zero was a formidable fighter. It was armed with two 7.7 mm machine guns and later with two 20 mm cannons. It was light and maneuverable. With a 925 hp radial engine, it had a top cruising speed of about 330 mph and could climb quickly to thirty-four thousand feet. To get such maneuverability, however, the designers skimped on armor plating and sacrificed protection for fuel tanks and pilots. (Bond and Anderson, *Flying Tiger's Diary*, 38)

Chennault emphasized that the Japanese Zero fighter was an exceptional airplane not to be taken lightly in combat. He stressed the Japanese airman was first rate and must be approached carefully. All of this said P-40s could hold their own in a dogfight if proper combat techniques were used by AVG airmen.

> The P-40 was much heavier, with a larger engine-an 1150 hp Allison-that was liquid cooled, meaning that a hit in the radiator could cause overheating and

Figure 6–2. Japanese Army Ki-27 Fixed-Gear Fighter. Main opponent of the AVG, known as the Nate. (Pacific Eagles, Nakajima Ki-27 Type 97 "Nate"))

> disaster. It had slower cruising speed, about 300 mph, and a ceiling of thirty-two thousand feet. (Bond and Anderson, *Flying Tiger's Diary*, 39)

In civilian talk, this means the Zero was faster and more maneuverable. These capabilities killed many unsuspecting Allied fliers from late 1941 until mid-1942. The P-40, while inferior in some ways, could defeat Japanese pilots when flown by competent, well-trained pilots. A Tomahawk (or any of the P-40 variants) when flown by pilots using Chennault's superior tactics consistently scored victories over the Oscar or Nate.

Design characteristics allowed the P-40 to accelerate in a descent Japanese aircraft could not match. When pilots were in trouble, they could dive and get away. The American engineered airframe was rugged and could sustain higher G forces than the more nimble Zero. Additionally, Curtiss-Wright designers had included an armored seat to protect the pilot. The P-40 used self-sealing gas tanks that theoretically assured greater combat survivability, and was armed with four 30- and two 50-caliber machine guns. With the heavy Allison engine mounted in the nose, the P-40 could dive faster than any Japanese airplane then in China. This was the flight characteristic that most kept the P-40 Tomahawk competitive.

Chennault preached that when all else failed the Curtiss Wright airplane was so strongly built it could ram Japanese aircraft and make it back to base. It sounds crazy but some P-40s actually rammed Japanese aircraft (sometimes on purpose) and survived.

Some Japanese aircraft outgunned the P-40 at short range (20- and 30-millimeter cannons are more destructive in close proximity than 30- or 50-caliber machine guns). The six P-40 machine guns had longer range and greater capacity for ammunition. The heavy 50-caliber slug easily shredded the thin aluminum skin of Japanese airplanes, and any internal structure it hit.

Many unsuspecting sons of Nippon died in 1941 and 1942 at the hands of well-trained AVG pilots. Chennault's airmen were told,

> Do not dogfight the agile Zero, avoid a turning fight with them. With a Zero on your tail, there is a good chance you might not survive. (Chennault, *Papers)*

All these things Chennault observed while flying his missions in China, because his Curtiss Hawk 75 had similar characteristics. A significant part of the Chennault mystique, this should be included in his legacy.

> The Old Man stressed these comparisons so that in combat we would make the best use of the strengths of our plane against the weaknesses of the Zero. The P-40 had the advantage in a dive and superior firepower; whereas the

Figure 6–3. Japanese Ki-43 Fighters in Formation. Known as an Oscar, this plane gave higher performance than the Nate. Such waves of attackers were a common view for the outnumbered AVG pilots. (Pacific Eagles, Kato Hayabusa Sentai-tai 1944, screen capture at 40.4 minutes)

> Zero could climb faster, turn sharper, and cruise higher. Chennault trained us to attain higher speeds by descending on the enemy, closing in, firing, and peeling off. Our greatest defensive strength in combat was the power dive. He also warned us to always fly in pairs-stick together. (Bond and Anderson, *Flying Tiger's Diary*, 39)

Correct application of Chennault's tactics allowed "inferior" P-40s to regularly defeat the aerodynamically superior Zero. Make the most of your strengths against the weaknesses of your opponent and you may live to play another day. Maxim Chennaultus!

Training accidents

Combat training is dangerous; accidents are always part of the equation. During training at Toungoo, the inevitable happened. There were crashes and casualties. The AVG lost their first planes and pilots. They buried their dead in the cemetery of St. Luke's, Church of England, in Toungoo.

Bright, energetic and popular Jack Armstrong of Hutchinson, Kansas, died in a mid-air collision. He was dog fighting with another P-40 when the accident occurred. Jack was an exceptional pilot. His loss stunned everyone. He was affable and well-liked, especially by the pilots. Until Armstrong's death, there was no real sense of danger associated with their flying. Knowing the risks incurred every day had a sobering effect on the AVG's true purpose, to kill Japanese.

Max Hammer from Cairo, Illinois, got lost in a rainstorm and while flying blind, crashed while trying to return to base. Although tragic, this type of accident was very common in the late 30s and early 40s. Instruments were primitive (non-existent in China) and when visual sighting was lost, pilots were at the mercy of God.

Pilot Pete Atkinson of Martinsburg, West Virginia, died after some sort of mechanical malfunction occurred in a dive to buzz the field. His airplane was virtually torn apart in flight. The motor fell free and dug a deep hole in the jungle on impact. Many thought this was due to a runaway propeller or that he exceeded design top speed of the P-40, but others felt the aircraft (previously crashed) just came apart. Either way, Pete was dead and Chennault lost another very good pilot.

Figure 6–4. Atkinson Crash. His P-40 disintegrated during a high-speed dive. The motor impact dug a deep hole in the ground. This photo shows a salvage attempt. (Neale, 230)

All of the fallen were honored with full military funerals conducted by Chaplain Paul Frillman, affectionately nicknamed "Holy Joe" by the pilots. Olga Greenlaw wrote to Armstrong's parents

> Dear Mr. and Mrs. Armstrong, Please don't thank me for the little things I did for John—who wouldn't have, knowing him—so fine and sweet? Thinking of John, I retain one consoling thought: He passed away quickly, painlessly. He never knew what happened. The collision stunned him and he became unconscious. The days that I knew him he was eager, expectant, and life held for him untold pleasures. I feel that this was so because he was one of the few boys that did not seem bored. He was so interested in everything about him, his training, the village, the Burmese people. He was so interested in my little dog Lucy, with whom he played every time he came to our house. He told me

> he had never seen such a funny looking dog in his life. His morals were the best. He was sufficient unto himself. His untimely death stunned all of us and made us realize that we were really at war and that it was a matter of life and death. To me and to all of us of the AVG, Armstrong is an unsung hero. And because of him, the Group "sobered up" and snapped to attention—"he made us all feel we had a duty to perform..." These were Jack Newkirk's words to me when we came back from his funeral. Now mother, do not grieve for this boy—he would not want it. Think of the AVG's score—John's score. (Letter from Olga Greenlaw to Mr. and Mrs. Armstrong, *1942)*[1]

Kindergarten, Chennault-style

Now it began in earnest, the education, discipline and determination that grew into the legend of the AVG.

R. T. Smith recalled Chennault this way:

> He was not a man to waste words and his manner was rather gruff but not unpleasant. He spoke in a low voice with a Louisiana cajun-country drawl that took a bit of getting used to before being completely understood. (Smith, *Tale of a Tiger*, 71)

Later that night as the men milled about the crew lounge, Chennault came in and sat down with his new flyers.

---

1 Author's note: I was attending a meeting in New Orleans in 2018. By coincidence, I ran into Jack Armstrong's nephew James Stahl. He told me of the family's 50-year search for Jack's body. Jack's parents wanted to return him to the family burial plot. Stahl traveled to Toungoo several times, trying to retrieve it. The Episcopal Church and Cemetery are long gone, destroyed by the Japanese. Finally, they learned the bodies of three AVG airmen were recovered by an allied burial team after the war. The remains were interred at the American Military Cemetery in the Punchbowl outside Honolulu, Hawaii. The remains were listed as unknown American airmen.

It took an additional twelve years to locate these graves. Each body needed DNA testing before they were positively identified. The bodies were returned to their families. Last year, Jack's remains were buried with full military honors and a hero's funeral. He was laid to rest beside his beloved mother and father. After seventy years, Jack was home again with his parents. Stahl's story overwhelmed me. I saw the amazement on this man's face when he learned of the additional information I uncovered while researching for this book. I told him of this letter and others between Olga Greenlaw and the Armstrong family. This information seemed to give a sense of peace to Stahl. God does move in mysterious ways!

> I will be giving lectures every morning on Japanese aerial tactics starting tomorrow at seven o'clock. I want to pass along one warning. Do not underrate the Jap pilots. They have had four years of combat experience. However, I think you and our P-40s will be a match for anything they throw at us. (Howard, *Roar of the Tiger*, 81)

There was still a long way to go until that happened, but from the first day onward there was no doubt what the end product would be. Chennault created the finest fighting air force America fielded in the Second World War.

> He lectured them on how to fight in pairs and how to engage in complicated maneuvers while staying together as a team, as perfectly coordinated as if both planes were being controlled by a single hand… He drilled them on the advantages of fighting in pairs… Fighting in pairs also allowed one pilot to concentrate on the target while the other protected his tail. (Schultz, *The Maverick War*, 115)

Why was the AVG so good?

The 1st American Volunteer Group was unmatched in effectiveness, maintaining a kill ratio over the enemy unequaled in air combat during the Second World War. Fighting seasoned Japanese airmen, inexperienced AVG pilots flew less capable aircraft against the finest fighter and bomber planes Japan produced. American airmen seemed outclassed in every aspect of combat. They weren't. The only recognizable difference was training. Chennault, the educator, ran a strict program. His so-called kindergarten made the difference.

> [F]rom the very first day he was enthusiastic about his work, his life's mission as it were. He would enjoy the tremendous opportunity to train a group of fighting men with complete freedom of action. (Samson, *The Flying Tiger*, 94–95)

Chennault's curriculum was made up of the Royal Air Force and stolen Japanese flight and staff manuals. He supplemented his personal observations on Japanese fighting tactics. These manuals were appropriated by his spy network. Translated and analyzed, AVG pilots learned as much about Japanese tactics as Japanese airmen. Understanding the enemy was an important part of Chennault's lesson plans.

Chennault's previous experience against the Japanese (it was rumored then and proven now) provided American volunteers valuable information about ways to engage successfully with their enemy. Tactics developed and taught by Chennault caused more destruction per aircraft available and produced greater negative impact upon the enemy than any commander in the history of aerial warfare, before or since. American and Allied pilots fighting in all combat theaters never approached the record of the AVG. It was a lucky day for the American volunteers and China when Chennault took command.

With Chennault in charge, the atmosphere around Toungoo changed. With a spring in his step and a growl in his voice, everyone believed Chennault would bring them to victory. He was that kind of leader.

> His words and his manner made a positive impression on most of the men. Chaplain Frillman noticed that within 24 hours, "the air of the camp had freshened." The men now had a leader and a sense of direction and purpose, all of which brought them closer to their objectives-adventure, a chance to fight the Japanese, and an opportunity to make good money doing it. (Schultz, *The Maverick War*, 110)

Managing an undisciplined and rowdy group of young men could be problematic for some. This was never the case with Chennault. He understood them, his command style was tailor made for them.

> My ideas on how to handle a group of high spirited, adventurous, volunteer fighter pilots and ground crews departed radically from military tradition. For rigid military discipline, I tried to substitute a measure of simple American democratic principle. Rigid discipline was confined to the air and combat matters. (Chennault, *Way of a Fighter*, 116)

Chennault's methods were very unusual for that period and were the perfect style of leadership for his pilots. He allowed everyone to participate in group meetings and everyone's opinion counted. Considering the wide range of personalities assembled, this helped maintain order and self discipline.

> There was never any formal discipline in the AVG camps, but the pilots always added a "sir" when addressing Chennault and called him "Colonel." Chennault was insistent on maintaining democracy within his group. (Hotz, *With General Chennault*, 117)

Figure 6–5. Pilot lounge at Toungoo. AVG pilots waiting for action. (Neale, 143)

His bearing demanded respect from all: everyone readily accepted Chennault's authority. Earned respect is a very good quality to have if you command men in combat.

Chennault was a strict taskmaster when he needed to be. He needed to be one now. According to Jim Howard,

> The Old Man set a 7:00 a.m. schedule for a lecture to all pilots. It was also a good time to take a roll call to make sure there were no malingerers. He also established a goal of sixty flight training hours, broken down into formation, tactics, gunnery, dogfighting, and cross country. *(Howard, Roar of the Tiger,* 88)

When it came to fighting, Chennault was very demanding. His pilots were novices with no combat experience. He knew the only thing between them and death was his training. If AVG pilots were to have a chance, he must school them in special techniques he learned while leading China's fight against the Japanese from 1937 until 1941.

> I never kept a record of his morning talks, but he had leaned heavily on his experiences in previous years as advisor to the Chinese Air Force. He never admitted to us that he had ever engaged in aerial combat, although it was rumored. (Howard, *Roar of the Tiger*, 88)

The educator in Chennault loved teaching his new pilots. What started as work was now his life's mission. He would:

> enjoy the tremendous opportunity to train a group of fighting men with complete freedom of action. He felt certain these would be the first American airmen to meet the Japanese; their lives were in his hands and he would largely determine how well they met the experienced Japanese flyers. (Samson, *The Flying Tiger*, 94–95)

Chennault recalled the first meeting with his new airmen. It was with a mix of humor and amazement divided into equal parts.

> Pilots looked far from promising as they checked in at Kyedaw. The long boat trip and Dutch shipboard menus had left many flabby and overweight. They all appeared wilted during their introduction to the humid monsoon heat. Field transportation was cut to a minimum to keep them walking and sweating; regular schedules of baseball, volleyball, and calisthenics were instituted to whip them back into good physical condition. (Chennault, *Way of a Fighter*, 110–111)

While getting his new charges into shape (mentally and physically) Chennault studied the personnel records of his men. In spite of his specific instructions, only a handful had any fighter experience.

> Their flying records were not that impressive. I wanted pilots between the ages of twenty-three and twenty-eight with at least three years of experience in fighter planes. Only a dozen met these standards and had ever seen a P-40. (Chennault, *Way of a Fighter*, 111)

In the end Chennault got what he wanted: a group of aggressive, intelligent, highly motivated young men. They were a mixture of military discipline with a whiff of a wild west mentality. These men were well suited for the combat style necessary to achieve their unlikely victory. They were adaptable and capable of handling difficult situations on their own. After absorbing Chennault's tactics they became a superb fighting force. These men were smart, skilled flyers who eagerly learned all they could about their enemy.

Once Chennault put his stamp of approval on each individual pilot they produced incredible results. Rough and ready they had enough grit to overcome any disadvantages and hardships dealt with in training or battle. Chennault took a group of untested young men and turned them into steely eyed killers. That sounds horrible but it was what they needed to survive. These men didn't just fight the vaunted Japanese war machine, they destroyed it. Their kill ratio exceeded all expectations, even their own. Exploits of these few American pilots were unequaled during the Second World War.

> Chennault had long cherished a volunteer force of American airmen, flying American equipment in China against the Jap. The purpose was fourfold: to test American equipment, to train a nucleus of American pilots in actual combat, to furnish air support to the Chinese land forces, and to fight a delaying action against the Japanese until the Chinese armies could be equipped with modern sinews of war for offensive action against the stranglehold of Japan. (Scott, *God is My Copilot*, 78)

Again the flaw in Japan's strategy reappears, dependence on rigid, unrealistic timetables for conquest. The Japanese High Command was lulled into complacency by the so-called "victory disease." Because they determined it should be so, they believed it would be so. But for Chennault it might have been so.

Knowledge as power, or the tactics of battle

Chennault supplemented stolen Japanese flight manuals with his own accumulated notes and conclusions. He flew many (combat) missions against the Japanese over the years. Knowing the minds of Japanese airmen and the tactics (predictable) they used in combat were a great help to the new, untested AVG pilots.

Personal experience taught him how the Japanese responded to tactical situations that arose. Chennault knew the flight characteristics and performance capabilities of every type of Japanese aircraft the AVG met. Knowledge is power in combat. With this knowledge Chennault trained his pilots how to best attack, where to shoot for quick destruction and how to safely disengage.

Chennault handed out mimeographed copies of specifications, performance data for Japanese airplanes and explained the best way for the P-40 to attack each type. These procedures remained on the blackboard until every pilot knew them like the back of his hand. The Colonel ended every lecture with the same words of caution,

> Forget the individual combat theory. Fight always in pairs. The lead ship goes in for the kill. The wingman protects his tail...It is the lead man who acts as the left hand, always pushing in, always probing until that short burst into the vitals of the enemy can be delivered. The wingman is the defender. He shifts from left to right, and then back again, always covering the lead ship to the flanks and behind. That way you make every thrust, every bullet count....Hit hard, go in with full power, everything to the firewall, but when you've made that attack with all determination, break away and get out. Go down with all the speed you have, for the heavy P-40 will take much more than the light Zero before you have to pull out. (Scott, *Flying Tiger*, 68)

These tactics are the same as those of J.E.B. Stuart in the American Civil War.

AVG pilots claimed Chennault devised the original combat procedure which allowed pilots flying in two-ship formations to cover each other. Weaving back and forth across each other's path both protected the other in a dogfight. This procedure later gained fame as the Thatch Weave, named for Naval aviator Jimmy Thatch. Chennault stressed the only acceptable surprises in aerial combat were those his men gave the Japanese.

There was a dual overt racism in the Pacific War. The Japanese were bigoted toward the (despised) white races while the Allies replied in kind with disdain for the Japanese. Both sides hated the other and did not consider their opponents as human. This battle would be brutal, bloody and cruel. How ironic, invading, benevolent Japanese warriors complaining about American cruelty. Bad, bad Americans!

Chennault made sure his pilots knew the strengths of the P-40 when compared against each type of Japanese aircraft. He emphasized how and where to attack. Chennault in the truest teacher fashion drew chalk airplane forms on the black board. He used this visual display to highlight where to do maximum damage to each type of aircraft. Chennault the educator taught his flyers the safest way to engage and how to get away when an attack went bad.

He explained different ways to conduct follow-on attacks. To achieve victory over superior Japanese aircraft he insisted they must be aggressive. Hit hard, fast, then make a clean high speed dive away. These were the cavalry tactics of Bedford Forrest or Jeb Stuart except now using aircraft. Chennault displayed this maneuver over and over to his fledgling fighter pilots. Eventually they could do it in their sleep. Each session of Kindergarten examined the advantages of the P-40 over all Japanese aircraft.

Chennault emphasized that AVG pilots must never get into a turning fight with more nimble Japanese airplanes. He explained their lightweight aluminum aircraft easily turned inside of the Americans. Within two turns a good Japanese pilot was in position to shoot them down.

> Never dogfight with Japanese airplanes; they are too maneuverable! Japanese [Zero] fighters out-climb the P-40 and it makes tighter turns. (Chennault, *Papers)*

Don't follow this advice and you will probably miss lunch today. He stressed to his flyers that Japanese planes have greater firepower with their cannon and machine guns so stay away from them whenever possible.

Weaknesses Chennault found with Japanese airplanes were revealed to AVG pilots. Their main shortcoming of Aircraft was the Japanese did not provide armored seats for their pilots' protection. Done since weight saving gave increased maneuverability, this proved a fatal flaw. They did not have self sealing fuel tanks either. Apparently Japan's pilots were not afraid of burning to death. By Japanese logic both pilot and airplane were expendable. Their aircraft were finesse machines and not as rugged as the P-40. Japanese planes flamed easily when hit in their gas tanks. Aircraft were of flimsy construction because of weight saving considerations that allowed greater range and speed.

When in trouble, P-40 pilots must perform a full power dive. Japanese aircraft were too light to stay with the Tomahawk (or any other P-40 variant) in a high-speed descent. Japanese planes were susceptible to machine gun fire (both 30- and 50-caliber). Aluminum shredded easily and often flamed when hit. Japanese airplanes were not as durable or rugged as the P-40.

> Remember when you are in trouble, dive away. Japs can't keep up. Once clear of the enemy just climb above them and begin another attack (Chennault, *Papers)*.

In terms of aircraft field maintenance and servicing, the Japanese aircraft were far superior to American designs.

> Japanese design philosophy was far advanced, intelligent and interesting... The Zero included many creative innovations. Its oil cooler, oil tank, American-type propeller and engine were assembled into one single unit held with only four big nuts to the plane's firewall [the front of the fuselage]. All fuel, oil, oil pressure, temperature and other service lines were connected into a single, simple, "quick disconnect" junction box.... To install or remove a complete Japanese power plant together with its propeller and oil cooler system took twenty-five to thirty minutes, while a similar job on its American counterpart, the Curtiss P-40 or newer North American P-51, needed five to eight hours! (Neumann, *Herman the German*, 99-100)

The Japanese were smart with their own engineering and open to stealing good ideas whenever they could. Their aircraft were responsive to the touch and more maneuverable than early American planes.

AVG P-40s were more rugged, had protective plating around the pilot seat, greater stability and a larger ammunition capacity. P-40s could dive like a bat out of hell. If all else failed, Chennault preached that the American plane could (and did) ram Japanese aircraft and fly away safely.

The AVG practices for war

Classroom discussions completed and the pilots began flying simulated combat attacks against one another. Chennault either flying above them or observing from a bamboo platform on the airfield, critiqued each pilot's attacks. Once the pilots landed Chennault debriefed each on his performance. It did not take long for the pilots to realize they were getting valuable information from a master pilot, one who learned his trade in combat. Lessons he identified from actual combat were taught in class. Chennault's instruction saved many lives in the coming days.

> When one considers that the AVG fought in what the British called obsolete combat aircraft-the P-40 B's and P-40 C's,-their deeds and scores become truly legendary. Throughout China today, Chennault's AVG are regarded as Saviors of Free China. (Scott, *God is my Copilot*, 79)

To the Chinese peasants AVG meant America Very Good, *ding-hao, ding-hao* (literal translation top good), America number one. It should be our goal to recapture these long lost memories of friendship. The Chinese were good friends once and now they are strong adversaries. Perhaps with mutual respect friendship might be rekindled.

Each new arrival trained in kindergarten. He was schooled in the Chennault way of combat. If a pilot needed more practice he was not allowed to fly in battle. Standardization of tactics was critical for evaluating progress made by his students. Only when a new pilot passed Chennault's criteria did they move on to the next stage of training. His kindergarten covered the basic necessities for fighting and surviving their upcoming Japanese experience.

> We began at Toungoo with a kindergarten for teaching bomber pilots how to fly fighters. Some learned fast and well. Bob Neale and David "Tex" Hill, both Navy dive-bomber pilots, had the best combat records in the A.V.G. George Burgard and Charley Bond, both Flying Fortress pilots, ranked among the first ten: for others it was a long, tedious, and unsatisfactory process. (Chennault, *Way of a Fighter*, 111)

Not all pilots have the same amount of skill or flying ability. Chennault patiently brought along each pilot at different rates until they possessed sufficient confidence "to safely fly with others."

Figure 6–6. Planning a Mission. Neale called this "Getting the dope from Chennault." (Neale, 136)

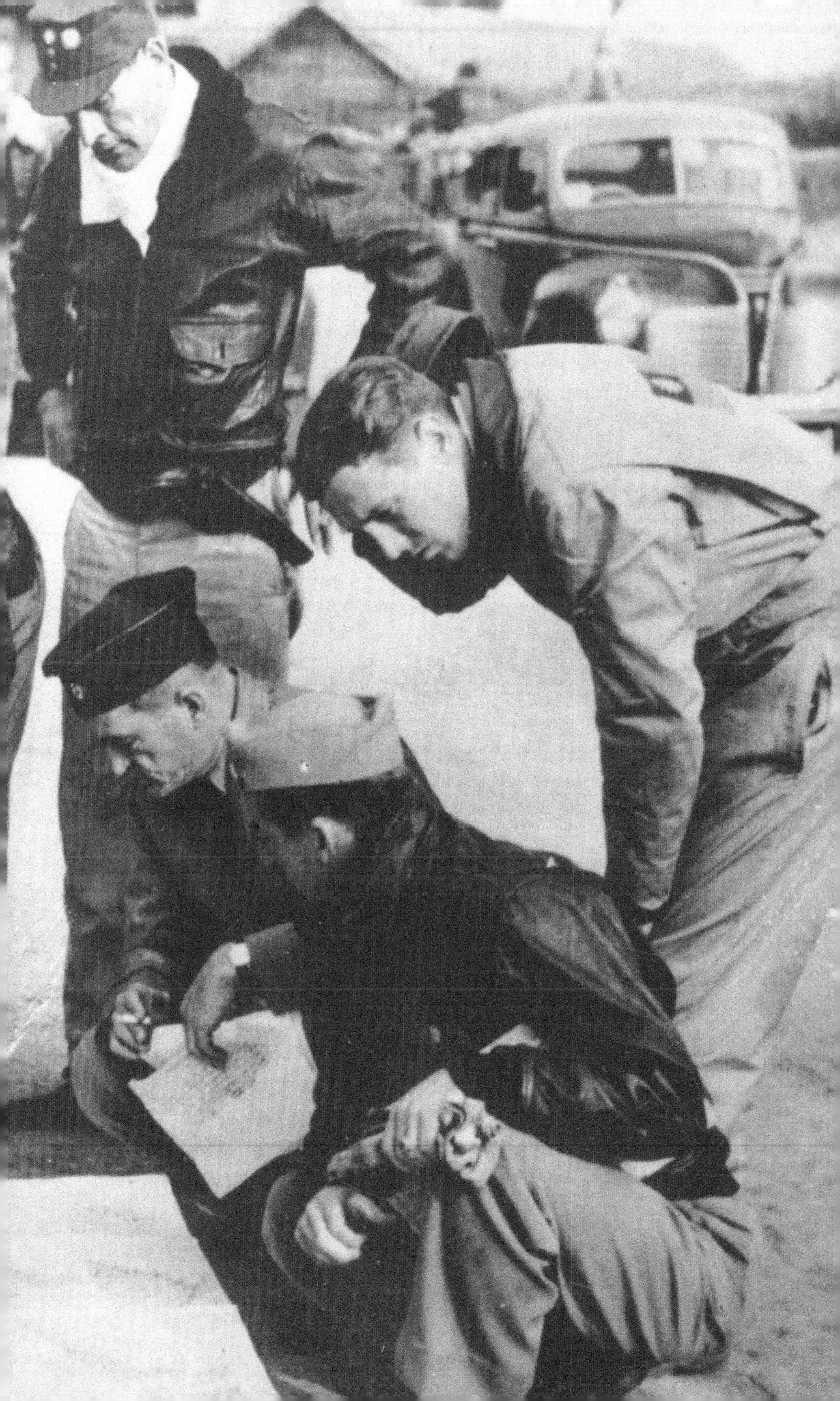

> Our training program went on long after combat began. As late as March 1942, after the group had been fighting for nearly four months, we had eighteen pilots classified as not ready for combat. No matter how pressing the immediate needs of combat I refused to throw a pilot into the fray until I was personally satisfied that he was properly trained. (Chennault, *Way of a Fighter*, 112)

The following excerpts come from the diary of a former Navy patrol bomber pilot while stationed at Kyedaw [Toungoo] airfield. A perfect example of Chennault's pilot upgrade program. It follows the progression from bomber pilot to fighter pilot with all the trials and tribulations.

> Oct. 31 – First hop in a P-40. Pretty tricky landing. They come in at 100 miles per hour. [Landing speed for his naval bomber approx. 85 miles per hour.]
>
> Nov. 4 – Tried stunts today. [Aerobatics] Almost got pretty sick. Not accustomed to such high speeds.
>
> Nov. 11 – First combat fight with Tex Hill. Pretty rough stuff-near blackout several times-guts heavy, neck sore. Tom Cole [another navy patrol bomber pilot] is disgusted with fighters.
>
> Nov. 12 – More combat practice. Now believe the old Navy stories that patrol duty is a pretty fine life. But I'll learn this stuff as well or better than any of the guys here. I still think I can fly anything. Tom Cole got on my tail in practice today. Now he likes fighters.
>
> Nov. 13 – Combat today with Cole. Jumping J—, that dogfighting is work! It will make an old man out of me—blacked out a couple of times. It's an awful feeling.
>
> Nov. 14 – Did pretty well in dogfight today with Ed Rector. Now satisfied I'll learn this stuff all right.
>
> Nov. 15 – Practiced interception problems and diving on bomber formations. Aerial gunnery too. Pretty fast pace these P-40's set. Know now it isn't too fast for me, though. (Hotz, *With General Chennault*, 109)

This fascinating passage walks us through a pilot's belief system about himself, his airplane and flying fighter combat. Chennault allowed each man to progress at his own pace. Here a pilot relates his progression from beginner (how do I fly this plane) to intermediate pilot (maybe I can fly a fighter) to being confident enough to go forward in this life or death scheme. Incredible, simply incredible!

> These early morning dogfights were weird affairs more spectacular than many of the actual combats that followed. The pilots unlimbered everything in the aerobatic repertoire, including a few tricks all of their own. Precision and teamwork was the theme. (Hotz, *With General Chennault*, 111)

Tactics taught here were the same Chennault gave his fighter pilot students at ACTS back in 1931. More proof of his visionary approach to combat fighter tactics! When you are right, you are right.

Chennault preached a disciplined attack by two-ship formations. When attacking, the wingman protects his leader. This tactic was later established as the best method for attack or defense. It was used by both American and Luftwaffe fighter pilots. Standard attack formations consisted of two two-ship formations flying together. The procedure is known as finger four or four-ship formation [the method is still in use today].

This configuration provides greater protection for both pilots. When entering combat it is the easiest way to split for attack. Chennault was a true pioneer of its combat utilization. His tactics were not revolutionary but they went to the heart of combat. Using hunting principles his father taught him as a boy, Chennault visualized the best method to attack.

> Just like hunting ducks, pick your target, be aggressive, proceed with confidence, kill your duck before he sees you. (Chennault, *Papers*)

Short and sweet right? Self-preservation and survival of friendly forces were the motivation for Chennault's tactics. His men loved him for it.

Figure 7–1. Disney Flying Tiger. (NicePNG)

7

The AVG Gets an Identity

A world away in Washington, DC, the principals of China Defense Supplies were considering a group logo for the AVG. After rejecting several ideas, T. V. Soong thought and

> then invoked the Chinese saying "Giving wings to the tiger." He reminded them that the tiger was the most formidable animal alive, but when endowed with wings its prowess was "super-colossal." Why not flying tigers? (Byrd, *Giving Wings to the Tiger*, 136)

So in October 1941, they asked the Disney studios to come up with a flying tiger logo. The Flying Tiger logo in various forms graced the AVG aircraft beginning in January 1942. It was adapted and extended for subsequent organizations and used throughout the war.

But no one consulted with the AVG. Back in Burma, the summer of 1941 wore into fall and men of the AVG developed a distinctive group personality, a cocky esprit de corps. Under Chennault, their morale was sky high. They were itching for a fight. These men wanted their enemy to know who they were. They designed their own branding, integrated into their planes and with the aggression they were soon to show to their enemies.

Sharkmouth

The P-40 had a pointed nose with a prominent air scoop underneath. One could easily see it as a gaping mouth. RAF pilots flying the P-40 in North Africa emphasized this.

> The shark-tooth insignia was not original with the AVG, although they made it famous. Three pilots, Lacy Mangleburg, Erik Shilling and George McMillan, picked up on the idea in Rangoon from a photograph in the *Illustrated India Weekly* showing Tomahawks of the RAF in Libya decorated with shark's teeth along the edge of the engine. (Hotz, *With General Chennault*, 106)

In mid-November of 1941,

> Chennault consented and mechanics set to work with paint brushes to convert the drab Tomahawks into leering tiger sharks. A teasing bloody tongue and a baleful red eye just behind the propeller were added to the design. (Hotz, *With General Chennault*, 106)

Once designed and painted on their airplanes, the Japanese would know who they were fighting. Using a brightly painted shark's mouth, the Japanese and the world soon learned about the AVG. However, they were not to become known as the Flying Sharks.

Figure 7-2. Painting a P-40. Inspired by a magazine article on a British unit in Africa, crew chief Van Timmeran paints the nose with shark teeth. The design of the P-40 accentuated the gaping mouth, becoming an unforgettable WWII icon. (Neale, 154)

FIGURE 7–3. ADAM AND EVE SQUADRON AT TOUNGOO. In addition to the shark mouth, the 1st Squadron's logo shows on the fuselage. (NEALE, 209)

With a group of aggressive, competent young fighting men like these, it was necessary to give them considerable freedom to express their combat persona. Several pilots began working on squadron insignia. Chennault recognized this and let his men decide how they decorated their personal aircraft.

> Theirs was a military organization that somehow was not military, and the men themselves were puzzled by it. One pilot despaired of discipline and was astounded at the constant griping, although "strangely enough, we do get things done, and in the last analysis it's the fight that counts." (Byrd, *Giving Wings to the Tiger*, 128)

Chennault showed organizational genius here. He was in charge of a civilian outfit made up of ex-military men fighting an undeclared war on an as yet undeclared enemy. Chennault realized he could not order them to do anything. You can't issue orders to a civilian; they just won't respond. This was an unprecedented situation and Chennault knew he must carefully handle these men. Ring knockers never understood then, some still don't.

The AVG was not a military formation, acting under accepted rules of war. A force unto themselves, the men made up the rules as they went along. No matter what each man's motivation when he signed up, they were mercenary soldiers of fortune now. Chennault and his men were making history; no one ever fought under these circumstances before. If successful in combat and not inducted into the U.S. military, this was the way they would remain. Even to the present day, most people don't understand this concept. Success is its own validation.

Men placed into difficult circumstances grow closer because of common experience. Hard times and shared danger make longtime friendships. Since their first days together, these men knew they were part of a special group. Many found comfort when they recognized old friends.

> Some of our former Navy pilots found others they had known at Pensacola or on duty with the fleet, and the ex-air corps were going through much the same process. (Smith, *Tale of the Tiger*, 60)

Like old home week, everyone knew someone from pre-AVG days. One benefit of military service in the early 1940s (because of the small size) was instant friendships. This added to the comradery, cohesion and esprit de corps the unit maintained from the first to the last day. Part of the myth of the AVG revolves around the respect its men had for each other. Friendships made quickly in dangerous circumstances lasted a lifetime.

Chennault faced a dilemma. He was in charge of a rowdy, aggressive, and sometimes undisciplined group of flyers from varying backgrounds. They were tough, not easily swayed, and discipline was certain to be a problem. Many pilots were high spirited, high energy types. They became sullen and reclusive if not handled just so.

Some turned to drink for stability, while others gave up drinking for the same reason. Men react differently to the same stimulation. It was important to maintain the correct motivation for each. Sometimes Chennault felt more like a ringmaster than a commander. No doubt about it, the men of the AVG were odd ducks. It took a special leader to keep them in line.

Style is everything when providing guidance to high-strung, aggressive young fighter pilot types. Too hard or too heavy a hand could damage these self-motivated men. Part of Chennault's "mystique" was his ability to take control of varied and sundry free spirits and mold them into the fighting team players necessary for his style of combat.

Jim Howard was one of Chennault's most trusted leaders, confidantes and friends. His book shows Chennault basing the Tigers' training on experience flying against the Japanese:

> The leadership method Chenault used with his squadrons was gradually becoming apparent. He was not flying any of the combat missions, yet he was commander of a flying organization. We knew, however, that he had flown combat missions against the Japanese before we came on the scene, so his advice to us on the performance characteristics of the Japanese planes was solidly based on experience. (Howard, *Roar of the Tiger*, 131–132)[1]

The question of Chennault's combat flying again rears its head. Howard believed Chennault flew scores of missions with many kills. If anyone knew from firsthand knowledge about Chennault, it was Jim Howard.

> The Colonel used to take me hunting with him, and I came to understand that throughout these hunting trips he was giving me lessons in tactics, lessons he had learned the hard way against the Japanese. Without my knowing it, he would, in effect, criticize my method of former attacks and advise me about better ways to do the job. I used to listen to him for hours as he told of cases in which he had got his ship shot up by going in too close, and then, after he learned how and knew that his longer range fifty caliber guns would outshoot the Jap, had accomplished the same destruction on the enemy without getting his own ship shot to pieces. These critiques taught me exactly what he meant to impart without ever hurting my pride by telling me that I was wrong and could accomplish more by fighting in his way. (Scott, *God Is My Copilot*, 219–220)

Chennault and military discipline

Out in the jungle, there was no Army brass and only Chennault made the rules. He had no regard for military regulations, needless distractions, or visiting dignitaries who disagreed with his system. You cannot command a volunteer to do anything: you must let him think you are asking him to do something. In some ways, these men were like children. You can show them the way, but they will only go

1 Author's note: First, this advice is pretty insightful for a man who purportedly never flew combat. In all of my research, this is the only acknowledgement from Chennault about his active combat flying. Second, it shows the quality of Chennault's leadership. Howard felt honored to serve under Chennault.

when they want. It was their war and they would fight it like they wanted, by God!

When the ring knockers returned, this was one reason cited for removing Chennault from his command. Regular Army officers saw many of the AVG members away from combat. They thought these men were slovenly, with disheveled uniforms and no discipline. These were not soldiers to be trusted, hell they probably weren't soldiers at all. This opinion gave no merit for their combat skills and did not bode well for their commanders' future.

Here is a prime example of why ring knockers didn't consider Chennault a successful military man. They had a long list of his flaws. He was insubordinate and always would be. He did not maintain discipline among his men, and his record keeping was atrocious. This opinion continued for the rest of his service career.

8

THE AVG PROVES ITS WORTH

The Burma Road was the only usable route between Rangoon and China, so this supply road was of critical importance. It was to protect the Burma Road that the AVG was originally organized and sent to China. If Japan closed this route, it meant the end of the Chinese as an Asian ally.

For the AVG at Toungoo:

> The three squadrons were shaping up nicely, and squadron commanders and flight leaders had been selected. In the mornings I'd stand with him on the control tower with a pad to write down notes of things he'd talk to the pilots about at the next briefing session. The Limeys finally gave us permission to have gunnery practice with our P-40s and pilot morale really improved with that. Practicing maneuvers and tactics was one thing, but knowing that your guns would shoot hard and straight was important to the airmen. (Trumble, *Biography)*

Then came the attack on Pearl Harbor. Japan had chosen plan Go-South to get oil. As part of the attacks that started with Pearl Harbor and attacks on the Philippines, Japan invaded both Thailand and the Malay peninsula on December 8, 1941. Throughout Malaya, the Japanese airforce with its highly maneuverable fighter planes simply annihilated the RAF, which clung to traditional dogfighting tactics.

FIGURE 8–1. ADAM AND EVE SQUADRON PREPARES TO DEPART. Two weeks after they painted the shark mouths, the 1st Squadron left Kyedaw for Kunming. (NEALE, 168)

Fast Japanese bombers flew unescorted and unmolested. While the British fortress of Singapore resisted, besieged. Thailand was rapidly occupied and could be used to attack Burma. The Japanese Army Air Force would be ready to use their long-range bombers to attack Rangoon by late December.

Chennault feared Kyedaw was very vulnerable to surprise attack, being just over the mountains on the border with Thailand. The warning net could not be extended into "neutral" Thailand, and there was no radar installed at Toungoo. He moved the First and Second Squadrons rapidly to Kunming. Chiang Kai-shek asked him to keep the Third Squadron at the main British airfield at Mingaladon, to assist the RAF in the defense of Rangoon and the Burma Road. Chennault insisted on maintaining operational control, so that the Hells Angel could continue to use his tactical approach to fighting the Japanese.

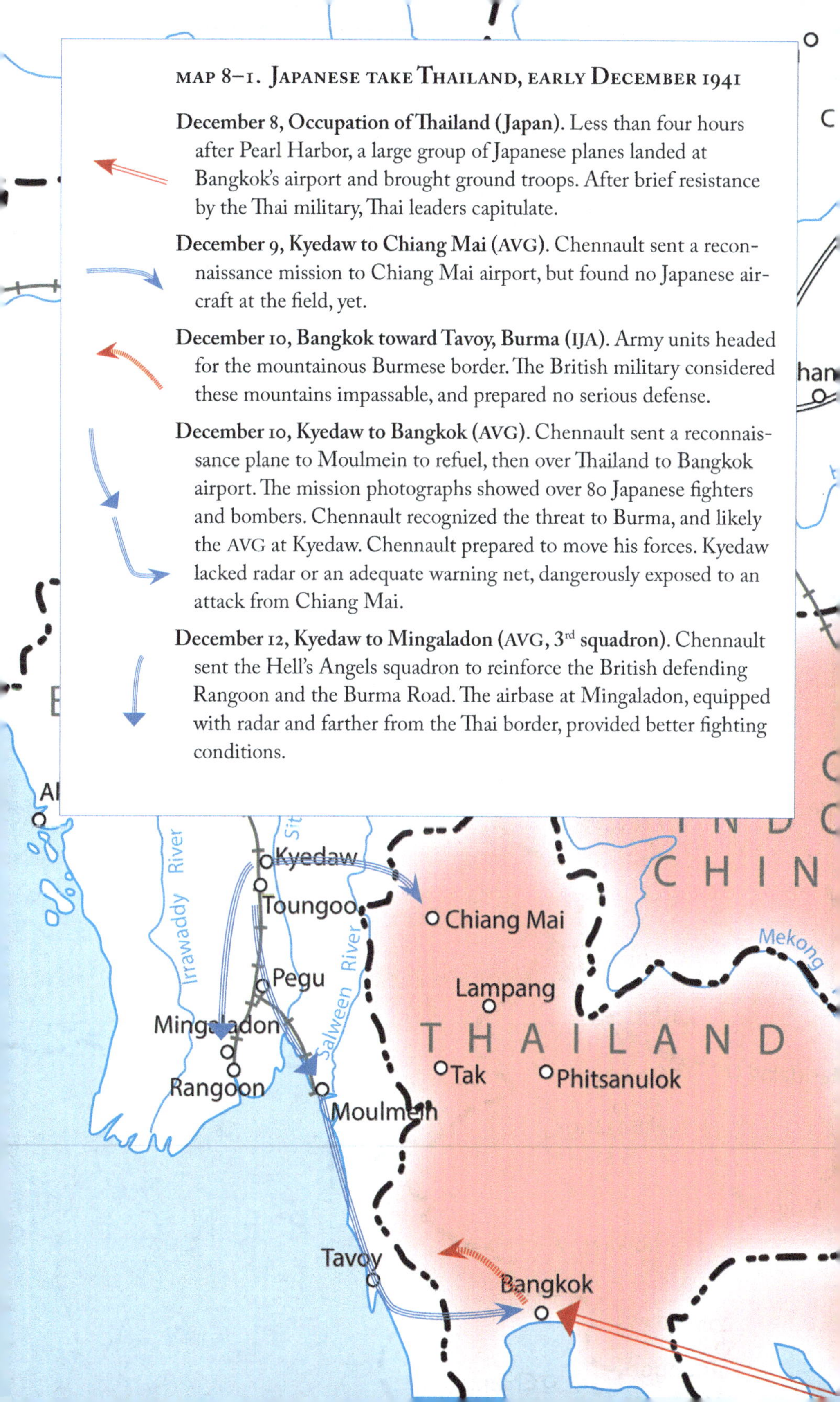

MAP 8–1. JAPANESE TAKE THAILAND, EARLY DECEMBER 1941

December 8, Occupation of Thailand (Japan). Less than four hours after Pearl Harbor, a large group of Japanese planes landed at Bangkok's airport and brought ground troops. After brief resistance by the Thai military, Thai leaders capitulate.

December 9, Kyedaw to Chiang Mai (AVG). Chennault sent a reconnaissance mission to Chiang Mai airport, but found no Japanese aircraft at the field, yet.

December 10, Bangkok toward Tavoy, Burma (IJA). Army units headed for the mountainous Burmese border. The British military considered these mountains impassable, and prepared no serious defense.

December 10, Kyedaw to Bangkok (AVG). Chennault sent a reconnaissance plane to Moulmein to refuel, then over Thailand to Bangkok airport. The mission photographs showed over 80 Japanese fighters and bombers. Chennault recognized the threat to Burma, and likely the AVG at Kyedaw. Chennault prepared to move his forces. Kyedaw lacked radar or an adequate warning net, dangerously exposed to an attack from Chiang Mai.

December 12, Kyedaw to Mingaladon (AVG, 3rd squadron). Chennault sent the Hell's Angels squadron to reinforce the British defending Rangoon and the Burma Road. The airbase at Mingaladon, equipped with radar and farther from the Thai border, provided better fighting conditions.

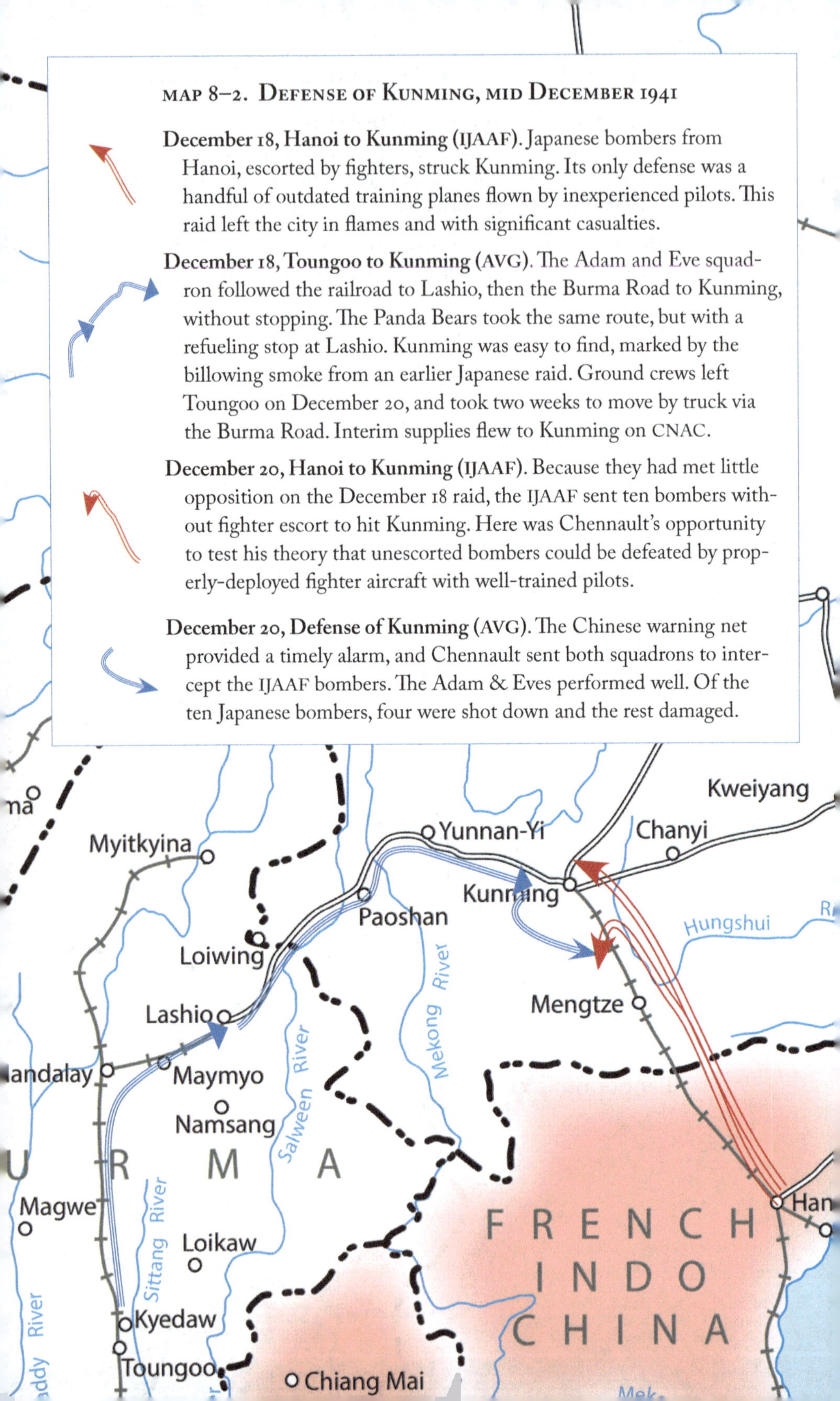

MAP 8–2. DEFENSE OF KUNMING, MID DECEMBER 1941

December 18, Hanoi to Kunming (IJAAF). Japanese bombers from Hanoi, escorted by fighters, struck Kunming. Its only defense was a handful of outdated training planes flown by inexperienced pilots. This raid left the city in flames and with significant casualties.

December 18, Toungoo to Kunming (AVG). The Adam and Eve squadron followed the railroad to Lashio, then the Burma Road to Kunming, without stopping. The Panda Bears took the same route, but with a refueling stop at Lashio. Kunming was easy to find, marked by the billowing smoke from an earlier Japanese raid. Ground crews left Toungoo on December 20, and took two weeks to move by truck via the Burma Road. Interim supplies flew to Kunming on CNAC.

December 20, Hanoi to Kunming (IJAAF). Because they had met little opposition on the December 18 raid, the IJAAF sent ten bombers without fighter escort to hit Kunming. Here was Chennault's opportunity to test his theory that unescorted bombers could be defeated by properly-deployed fighter aircraft with well-trained pilots.

December 20, Defense of Kunming (AVG). The Chinese warning net provided a timely alarm, and Chennault sent both squadrons to intercept the IJAAF bombers. The Adam & Eves performed well. Of the ten Japanese bombers, four were shot down and the rest damaged.

First battle of the AVG

Meanwhile, a Japanese bomber squadron in transit to Malaya arrived in Hanoi. They conducted a bombing raid on Kunming on December 18, 1941. Chennault's fighters arrived after the raid to find the city in flames. Two days later, the Japanese tried again, but without escort because they had had no fighter opposition. The warning net gave ample time for the AVG to intercept. Chennault waited in the fortified combat-operations center dugout as reports of the approaching bombers were plotted.

> This was the decisive moment I had been awaiting for more than four years—American pilots in American fighter planes aided by a Chinese ground warning net about to tackle a formation of the Imperial Japanese Air Force, which was then sweeping the Pacific skies victorious everywhere. I felt that the fate of China was riding in the P-40 cockpits through the wintery sky over Yunnan. I yearned to heartily to be ten years younger and crouched in a cockpit instead of a dugout, tasting the stale rubber of an oxygen mask and peering ahead into the limitless space through the cherry-red rings of a gunsight. (Chennault, *Way of a Fighter*, 128-129)

At least four of the ten bombers were promptly shot down. Reports vary as to how many were able to return to base damaged, but no more unescorted raids on Kunming occurred.

> The shooting began rather earlier than Chennault and his 100 boys with their 54 obsolescent planes had bargained for. But the account which the A.V.G. then proceeded to give of itself justified not only Chennault's untried gospel of the flying team but America's untried faith in her sons of the sky. (Author, *LIFE*, July 20, 1942 issue, 4)

December Rangoon bombing raids

By late December, the Japanese Army Air Force assembled in Thailand enough aircraft to launch a devastating air raid on Rangoon. Flying unescorted from bases in southern Thailand, sixty heavy bombers targeted Rangoon

MAP 8–3. DEFENSE OF RANGOON, LATE DECEMBER 1941

The IJAAF staged two large bombing raids on the city of Rangoon and Mingaladon air base. These two raids successfully bombed both targets, but at considerable cost in bombers lost to AVG guns. Attack routes were the same for both December 23 and 25.

Northern Group to Mingaladon (IJAAF). Fighters from forward bases and older bombers from more distant bases in northern Thailand attacked Mingaladon.

Southern Group to Rangoon (IJAAF). Heavy bombers from bases in southern Thailand attacked Rangoon, generally without fighter escort.

British defense. British Buffalos based at Mingaladon attack the Japanese bombers. Some are lost to the Japanese fighter escorts.

AVG defense. AVG P-40s attacked in small groups, using their high diving speed to great advantage. Enough bombers were shot down that Japanese suspend daylight raids.

Continued advance on Tavoy, Burma (IJA). Undetected in the thick jungle, Japanese Army units continued across the mountainous Burmese border.

New advance on Moulmein, Burma (IJA). Another Japanese Army unit started through the mountainous Burmese border towards Moulmein.

and the war material being unloaded at its docks. Flying from bases in northern Thailand and targeting the RAF airbase at Mingaladon was a smaller force: only fifteen heavy bombers, twenty-eight light bombers and thirty-two fighters as escorts. The plan was to arrive over both targets simultaneously, overwhelm the RAF and AVG fighter defense, and repeat as necessary. For comparison, the AVG had fourteen P-40s and the RAF had sixteen Buffalos operational.

The first large Japanese raid on Rangoon and its Mingaladon airfield happened on December 23, 1941, but not as the Japanese planned it. The two northern groups of bombers targeting Mingaladon became separated, with the heavy bombers well out in front of the rest. The RAF detected the incoming raid by radar but failed to alert the AVG for half an hour. Still the AVG got airborne in small groups as they could and successfully intercepted the heavy bombers. The AVG and the RAF shot down five heavy bombers, a third of the attacking force!

The southern IJAAF force of heavy bombers also became separated over Bangkok and never regrouped. The first group of eighteen was intercepted by five P-40s of the AVG, but continued on to drop their bombs on Rangoon. Two were shot down and another badly damaged with the group commander killed. The second group arrived ten minutes later and were not intercepted. Twenty-seven heavy bombers each carrying 2,200 lbs of bombs devastated Rangoon.

The Japanese sent another raid on December 25, and this time the British radar provided the AVG adequate warning. Ground crews had fixed various P-40s so that thirteen were ready to fly. Again some of the IJAAF heavy bomber groups became separated from their fighter escort, and five heavy bombers were shot down. The AVG lost none and the RAF four of fourteen Buffalos. Damage to Rangoon and Mingaladon was heavy.

> On Dec. 23rd, and again on Christmas Day, the Japanese Air Force appeared in awesome numbers; some 60 twin-engine bombers and 30 fighters came swarming from bases in Thailand. We met them with 14 P-40s, all we had available, and in the two encounters, shot down a total of 35 Jap bombers and fighters, damaged many more, with the loss of five P-40s and two pilots. This, my friends, was the ultimate in on-the-job training! (RT Smith, *History of the Flying Tigers*, usshawkbill.com)

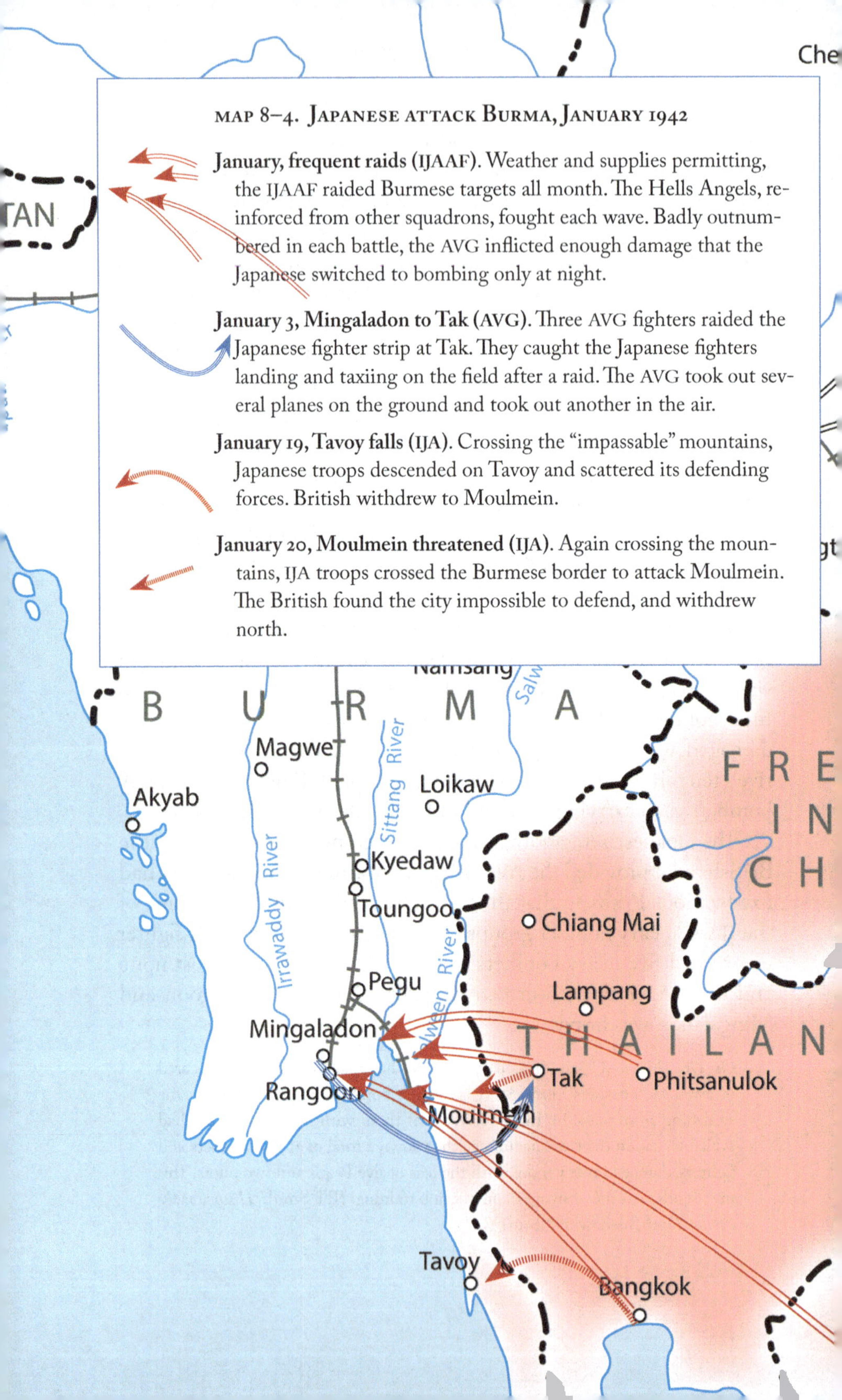

MAP 8–4. JAPANESE ATTACK BURMA, JANUARY 1942

January, frequent raids (IJAAF). Weather and supplies permitting, the IJAAF raided Burmese targets all month. The Hells Angels, reinforced from other squadrons, fought each wave. Badly outnumbered in each battle, the AVG inflicted enough damage that the Japanese switched to bombing only at night.

January 3, Mingaladon to Tak (AVG). Three AVG fighters raided the Japanese fighter strip at Tak. They caught the Japanese fighters landing and taxiing on the field after a raid. The AVG took out several planes on the ground and took out another in the air.

January 19, Tavoy falls (IJA). Crossing the "impassable" mountains, Japanese troops descended on Tavoy and scattered its defending forces. British withdrew to Moulmein.

January 20, Moulmein threatened (IJA). Again crossing the mountains, IJA troops crossed the Burmese border to attack Moulmein. The British found the city impossible to defend, and withdrew north.

When he was in the Army Air Corps, Chennault vociferously disagreed with the prevailing "the bomber will always get through" theory, to the detriment of his career. Here the bombers had gotten through, but at a prohibitive price. In two days, twelve of the seventy-one IJAAF heavy bombers had been lost, and more damaged.

The Japanese decided to hold back further large bomber raids on Rangoon, using the heavy bombers instead in Malaya and Singapore where there was no effective fighter opposition. The light bombers were retained in Thailand where they conducted only night raids on Mingaladon. Thus there was a lull in the air war over Burma.

Chennault would later say of his pilots

> Although the AVG was blooded over China, it was the air battles over Rangoon that stamped the hallmark of its fame as the Flying Tigers. (Chennault, *Way of a Fighter*, 130)

Action in Kunming

For almost a month, no Japanese bomber had threatened Kunming. The AVG squadrons there had little to do, so Chennault planned a raid on Hanoi. At the same time, a Japanese heavy bomber group was staging through Hanoi, and decided to send three heavy bombers towards Kunming on January 17. Their target was Mengtze, site of a former CAF training base along the rail line between the two cities (see Map 8–2). Also in Mengtze was a mine that was a major source of tungsten for the United States.

The fighter escort for the three bombers got lost, but the bombers continued unescorted. The radio station at the former training base was able to direct the AVG to the incoming bombers, in those days a novelty that Chennault had long advocated. Two of the bombers were shot down, and the third damaged too badly to return to Hanoi. Here again, Chennault's theories of defensive pursuit were vindicated.

Attacking Japanese air bases in Thailand

To make up for losses in Burma, Chennault sent eight additional planes from the 1st Squadron (Adam and Eve) at Kunming to Mingaladon airfield at Rangoon. With reinforcements in place and a break in the air war, Chennault initiated the first AVG strafing attacks on Japanese air bases in Thailand. Surprise was the only advan-

Figure 8–2. Peter Wright Crash, Mingaladon. For night landings, cars were parked on the sides to illuminate the runway. After scrambling to intercept a Japanese night raid, AVG pilot Wright landed slightly askew, the gear collapsed and the plane slid into one of the cars. Another pilot, Ken Merritt, was instantly killed when the sliding plane hit the vehicle he was in. (Neale, 226)

tage, since these bases had anti-aircraft guns. Starting an attack by diving aligned with the rising sun worked, but only once. Circling for more attacks until all ammunition was gone greatly increased a pilot's chance of being hit by ground fire. Strafing was dangerous business—Mott was forced down and captured Jan 8, and Moss was hit and bailed out over the mountains but managed to walk out to Burma.

Japanese troops advance into Burma

Although there was a lull in the air war, the same was not true on the ground. Japanese troops began hacking their way through the jungles and mountains separating Burma from Thailand in two places in January. The first group to emerge attacked Tavoy on January 19, routing the local defense forces and seizing the airfield.

British forces retreated rapidly to Moulmein along the rail line, destroying the bridges as they went. However, Japanese troops advanced rapidly using bicycles. They used the destroyed bridges as supports for makeshift wooden bridges and continued to advance, just as they were doing as they advanced the length of Malaya toward Singapore.

Figure 8–3. Japanese Bypassing Demolished Bridge. Because British troops depended on trucks to travel, a blown bridge could stop them until repaired or bypassed. So they misjudged how quickly Japanese troops might lash together a foot bridge and continue. Here IJA troops take their bicycles across a river south of Moulmein. (Fold3, 5306000)

January Rangoon bombing raids

The next Japanese air attacks occurred late in January, but then became daily events. A new group of nineteen heavy bombers had arrived, no longer needed over Hong Kong or the Philippines. Their commander tried daylight bombing again, this time using a limited number of their heavy bombers protected by many more fighters.

> By the last week in January, the Japanese were ready for another knockout attempt on Rangoon. From January 23 to 28 six major attacks of up to one hundred planes each rolled over the Burmese port. It was a tribute to the Anglo-American fighter pilots that the Japanese formations had switched to a three-to-one ratio of fighters protecting small bomber formations. (Chennault, *Way of a Fighter*, 135)

Burma was a killing ground for each AVG squadron as they rotated into combat. The newly named Flying Tigers overwhelmed numerically superior Japanese air forces most of the time, but not always without cost to themselves. No matter how many enemy aircraft attacked, regardless of the leader, the kill ratio heavily favored the AVG. Chennault's training and instruction were spot-on. Japan did not know it yet, but American pilots flying outdated airplanes were outclassing them. While Burma proved an AVG killing ground, the killing went both ways. With increased AVG combat operations, it was inevitable that more Americans would die. Some were in combat, but accidents claimed pilots and planes too.

> Saturday, January 31st, 1942 – AVG pilots lost in January: Jan. 8th.-K. Merritt, accident. Mingaladon, Burma. Jan. 8th,-C.D. Mott, missing in action-Bangkok, Thailand. (Probably alive and captured by the enemy). Jan 23rd, *Christman*, in action Rangoon, Burma. Jan 26th,-Hoffman, in action, Rangoon, Burma. Jan 20th,-T.J. Cole, in action, Rangoon, Burma. (AVG War Diary)

The objective of these Japanese attacks was the AVG at Mingaladon. The first day the bombers couldn't rendezvous with the fighters, so the fighters went on alone and lost some to AVG guns. The second day, January 24, the six heavy bombers outpaced their escorts again and all were shot down by the defenders, nine fighters of the RAF and AVG. They switched back to night raids, and fighter-only sweeps to win a war of attrition.

Figure 8–4. Salvaging Guns From a Downed P-40. With replacement parts often unavailable, crews scavenged for any usable parts after a crash. (Neale, 171)

> On January 23 and 24 the Japanese tried to floor the A.V.G. with a series of one-two punches. They led with a fighter sweep designed to get the Allied fighters into the air and use up their fuel. Then a second wave was scheduled to deliver the knockout punch while the A.V.G. and R.A.F. were on the ground refueling. It was a good plan but it didn't work. A.V.G. ground crews were too fast on refueling and rearming the P-40's and had them ready to fight again before the second wave of Japs appeared. By January 28 the Japs were sending over only large fighter formations, and the score for this offensive stood at 50 Jap planes destroyed against a loss of 2 A.V.G. pilots and 10 R.A.F. pilots killed. (Chennault, *Way of a Fighter*, 135)

Supplies, always those damned supplies

Chennault's most pressing problem was not Japanese airplanes but Allied supply lines. The AVG was short of everything needed to put up a good fight. Frayed, short tempers were in abundance, with no way to allay the frustration everyone felt. Pilots were exhausted, as were their airplanes. Combat was wearing on the pilots and it showed. Olga Greenlaw observed the changes,

> Almost every member of the group was beginning to show signs of war weariness. Pilots and ground crews had taken a terrific pounding and their nerves were starting to fray along the edges. Wear and tear on the planes was enormous and our general situation was growing steadily worse. We didn't have enough of anything-plane replacements, spare parts, and anti-aircraft weapons. (Schultz, *The Maverick War*, 202–203)

Some replacement aircraft were in the pipeline, but they were not being rushed to the front.

> Regarding the P-40s and P-43's which have been detained in Calcutta by bad weather for some time, it is suggested that they proceed to Dinjan and act as convoy for U.S. Ferry Planes over the route Dinjan-Myitkinya-Yunnan-Yi. The pilots of the Ferry Planes would be very glad to have them, and at the same time would guide them into Yunnan-Yi or Kunming safely. (Harvey Greenlaw, *Request for aircraft to Material Command/Karachi*, India February 1942)

For the AVG, life at Mingaladon got worse with each passing day. The Japanese bombed regularly and Chennault's men lost equipment and ate up supplies faster than they could be replaced. P-40s already exceeded their expected engine life. After each mission, there were fewer and fewer planes available to fly. Those that could were barely maintained.

> We are badly in need of these overhaul tools and it is requested that urgent efforts be made to have them shipped here by U.S. Ferry Command, Pan American, or CNAC. We must have tools and we want them laid down at Kunming for immediate use. (Harvey Greenlaw, *Requisition to Material Support Department*, Karachi, India, late January 1942)

Supply, or its lack, remained the curse of the AVG. Shortages of material and aircraft were Chennault's Achilles Heel. Japs they could kill, at least until they ran out of bullets and airplanes. Airplanes don't fly without maintenance, no tools-no fix-no can fly. It was just that simple. Supply Command in India was more destructive to the AVG than the Japanese flyers they faced in aerial warfare. What a way to fight a war!

9

The Japanese Take Burma

Singapore and Malaya

The Japanese army moved rapidly south through Malaya towards Singapore, the British Gibraltar of the East. British efforts to slow the advance (destroying bridges, etc) did little in the face of Japanese ingenuity and familiarity with jungle warfare. The Japanese air force used their superior fighters and tactics to destroy the RAF there.

The "impossible" fall of Singapore happened on Saturday, February 13, 1942. Now Japanese forces could concentrate on attacking Burma. This did not portend well for Chennault and the AVG. They were next in line to face the full might of the vaunted Japanese.

> After the fall of Singapore in mid-February, the Japanese transferred the crack air units that blasted the R.A.F. out of the Malayan air to Thailand to join the assault on Rangoon. (Chennault, *Way of a Fighter*, 137)

The Japanese juggernaut appeared unstoppable. In the five weeks since Pearl Harbor crippled the American fleet, the Japanese had seized French Indochina, Thailand, Malaya and the islands of Singapore, Wake and Hong Kong. The American air forces in the Philippines had been wiped out. Most of the Philippine Island of Luzon including Manila were now under Japanese control.

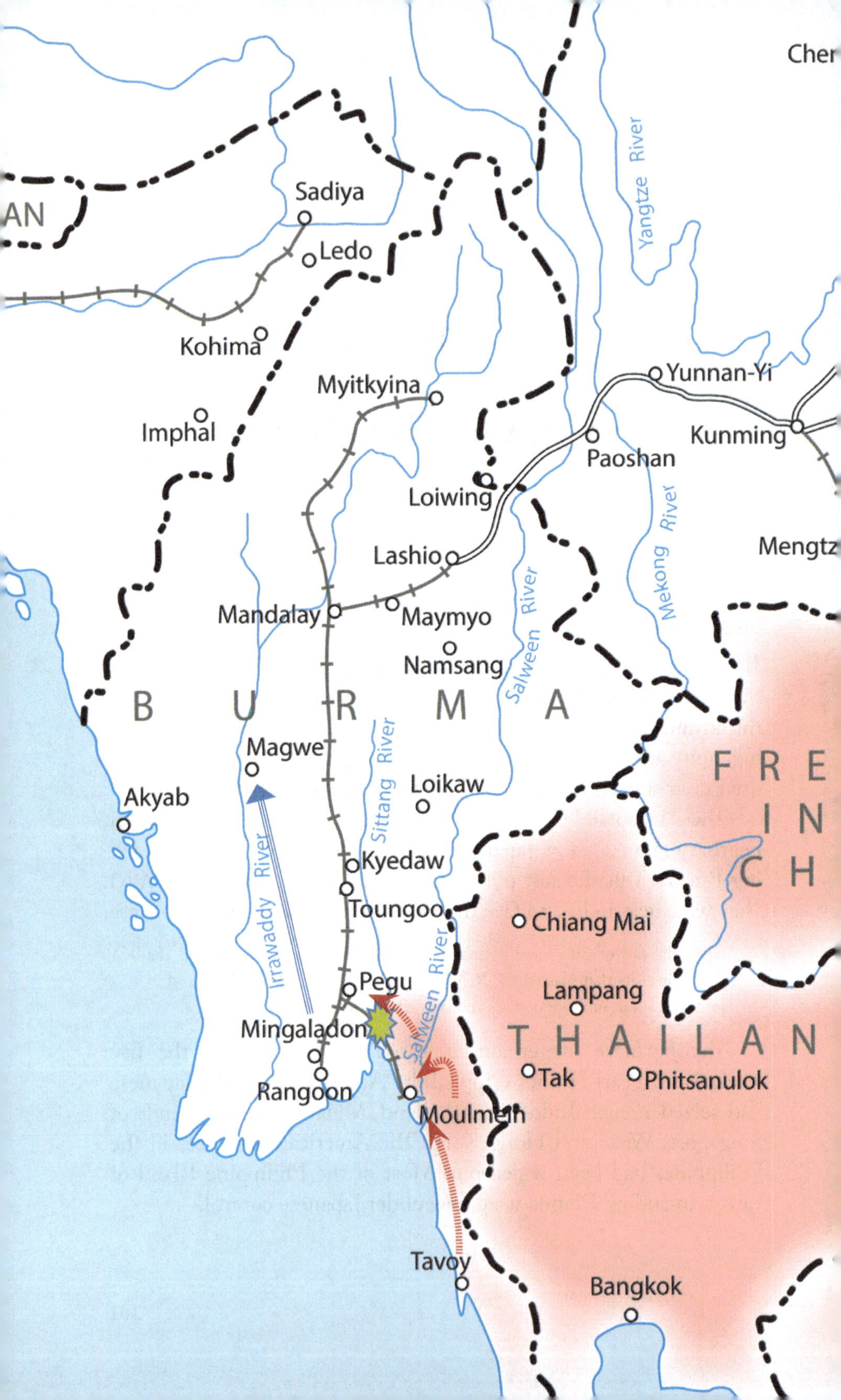

Chen
AN
Sadiya
Ledo
Kohima
Imphal
Myitkyina
Yangtze River
Yunnan-Yi
Kunming
Paoshan
Loiwing
Lashio
Mekong River
Mengtz
Mandalay
Maymyo
Namsang
Salween River
BURMA
FRENCH
Magwe
Sittang River
Loikaw
Akyab
Irrawaddy River
Kyedaw
Toungoo
Chiang Mai
Pegu
Salween River
Lampang
Mingaladon
THAILAN
Rangoon
Tak
Phitsanulok
Moulmein
Tavoy
Bangkok

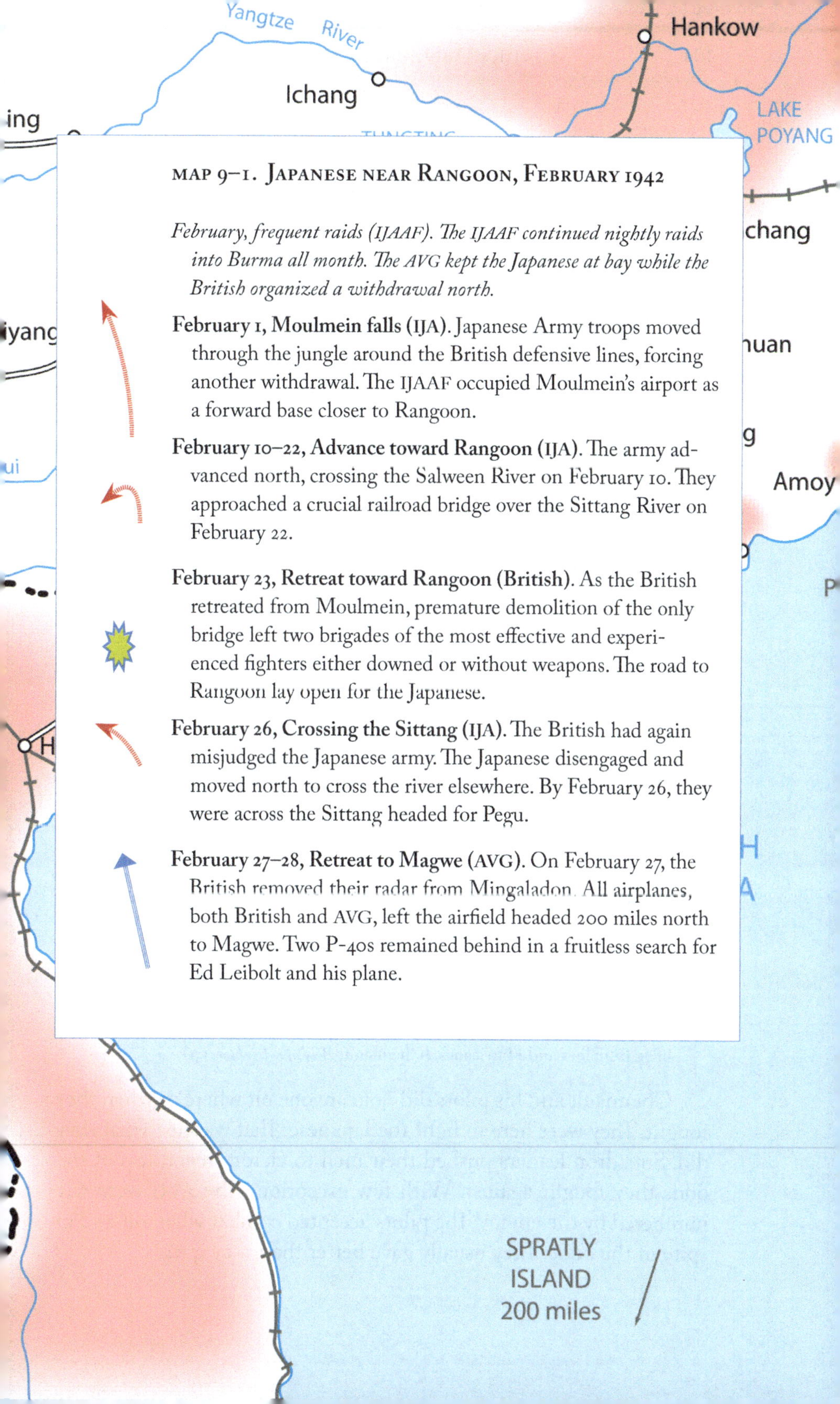

MAP 9–1. JAPANESE NEAR RANGOON, FEBRUARY 1942

February, frequent raids (IJAAF). The IJAAF continued nightly raids into Burma all month. The AVG kept the Japanese at bay while the British organized a withdrawal north.

February 1, Moulmein falls (IJA). Japanese Army troops moved through the jungle around the British defensive lines, forcing another withdrawal. The IJAAF occupied Moulmein's airport as a forward base closer to Rangoon.

February 10–22, Advance toward Rangoon (IJA). The army advanced north, crossing the Salween River on February 10. They approached a crucial railroad bridge over the Sittang River on February 22.

February 23, Retreat toward Rangoon (British). As the British retreated from Moulmein, premature demolition of the only bridge left two brigades of the most effective and experienced fighters either downed or without weapons. The road to Rangoon lay open for the Japanese.

February 26, Crossing the Sittang (IJA). The British had again misjudged the Japanese army. The Japanese disengaged and moved north to cross the river elsewhere. By February 26, they were across the Sittang headed for Pegu.

February 27–28, Retreat to Magwe (AVG). On February 27, the British removed their radar from Mingaladon. All airplanes, both British and AVG, left the airfield headed 200 miles north to Magwe. Two P-40s remained behind in a fruitless search for Ed Leibolt and his plane.

> To contrast this depressing news, the victories of the AVG sounded the only encouraging note. And so our squadrons at Mingaladon continued to be visited by newsmen and photographers. (Howard, *Roar of the Tiger*, 118)

A bleak picture for sure, and many more defeats would soon befall the overwhelmed Allies worldwide. In darkness a light, any light, shines brightest. That light in this instance was the Flying Tigers. "We raised all kinds of hell on a shoestring, said one AVG boy," *Life Magazine* reported on July 20, 1942.

Promising news for the Americans and their Allies was desperately needed during the opening phase of the war. The only positive actions then being fought were in the skies over Burma and China. These victories were enthusiastically credited to the Flying Tigers and their brilliant commander Claire Lee Chennault by journalists, eager for good news. Whatever the exact numbers, the Tigers' skill, daring-do and training resulted in the first effective defense against Japan's fearsome air force, a morale boost for Allied forces.

February Rangoon bombing raids

The AVG remained in combat in Burma as long as the Japanese attempted to bomb Rangoon and Mingaladon, which was nearly every night. Conditions they operated under were primitive with a lack of everything necessary for effective fighting. Supply for the AVG was non-existent, forcing Chennault to scrounge or recycle everything possible (including wrecked airplanes).

In spite of all these handicaps, the AVG maintained an outstanding combat record.

> In thirty-one encounters, they destroyed 217 enemy planes and probably destroyed 43. Our losses in combat were four pilots killed....Sixteen P-40s were destroyed. During the same period, the RAF, fighting side by side with the AVG, destroyed 74 enemy planes, probably destroying another 33, with a loss of 22 Buffaloes and Hurricanes. (Chennault, *Way of a Fighter*, 131)

Chennault and his pilots did not care one bit where or when they fought. They were here to fight the Japanese. That was just what they did. Squadron leaders pushed their men to victory regardless of the odds they fought against. With few exceptions, the AVG were outnumbered by the enemy. The pilots accepted combat when offered in spite of the odds. They usually gave better than they got.

> One shining hope has emerged from three catastrophic months of war. That is the American Volunteer Group pilots, the so-called "Flying Tigers" of Burma and southeast China, who paint the jaws of a shark on their Curtis P-40's. Outnumbered often ten to one, they have so far shot down about 300 Japs planes, killed perhaps 800 Jap airmen. They have violently wretched from the Jap Air Force control of the skies over Burma and southeast China. They have conclusively proved what once was only a Yankee belief: that one American flyer is equal to two or three Japs. "Give me," said U.S. Lieutenant General Brett in Australia last week, "100 fighters to 200 Japs and I'll lick them every time. And I'm not disparaging the Japs. They are good fighters." (Life Magazine, *30* March, 1942)

Recognizing this outstanding effort, Winston Churchill, Prime Minister of the United Kingdom, said,

> The victories of these Americans over the rice paddies of Burma are comparable in character, if not in scope, with those won by the RAF over the hop fields of Kent in the Battle of Britain. (Chennault, *Way of a Fighter*, 131)

The kill ratio of British to German planes in the Battle of Britain was one to four, an exceptional achievement. The kill ratio of Anglo-American fighters to Japanese planes over Rangoon was one to fourteen, an unbelievable achievement. This feat is unequaled by any air force anywhere, either before or after the Burma air campaign.

> Using the tactics of diving upon them from above with deadly gunnery, the P40s literally tore the Zeros and bombers apart in diving speed which they couldn't match and then climbing swiftly to circle above and attack again. The RAF Brewster-Buffaloes also stationed at Mingaladon had neither the tactics nor the fast diving abilities of the P40 and their Zeroes with incredible ability to turn and maneuver in short distance took a heavy toll of courageous Royal Air Force pilots. (Trumble, *Biography)*

The Japanese maintained aerial dominance throughout the Pacific. Chennault's Americans continued flying their worn-out P-40s. These comprised the only Allied air forces left in Asia. Chennault's men now worked closely with British flyers.

Attitudes on both sides began to change. Americans originally found the British standoffish, stuffy and snobbish. British pilots thought Americans were cowboys out of the wild west like in the movies. Pistols on their hips, they seemed unpredictable and undisciplined. Working and flying side by side revealed the true courage, determination and professionalism of both groups. Now considered brothers in combat they gave the highest accolade one flier can give to another. These men flew and fought monumental battles and neither side ever worried about who was covering his wing.

In the rush of combat, all a pilot thinks about is putting himself into the proper firing position. This, more than anything else, determines success. Properly positioned, the attacker has all the advantages. He picks when and where to attack for maximum effect. In combat, the pilot feels the thrill of victory or the agony of defeat. War is a dangerous business with no such thing as second place.

Disaster for the British on the ground

As the British retreated from Moulmein, the only escape north to Rangoon was the railroad bridge at the mouth of the Sittang River. They planned to destroy the bridge after crossing, to stop the Japanese, approaching rapidly from the south. A tragic mistake led to its premature demolition, trapping a significant portion of the British army on the east bank. Leaving their equipment onshore, many drowned while attempting to swim across. This British debacle prevented any significant defense of the west bank, as half of the force, two brigades of the most effective and experienced fighters, had either drowned or had no weapons. The road to Rangoon lay open.

AVG and RAF withdraw from Rangoon

Chennault could do nothing; the message of the prophet was already written on the wall. Death was coming to town. All that remained for him was to decide when and how to withdraw the Tigers from Rangoon. He weighed all of his options before making his final decision. Whenever the AVG withdrew, Chennault knew it was the end of the road for Rangoon.

> It was during this period that a handful of battered P-40's flown by Bob Neale's First Squadron pilots wrote the final lurid chapter in the A.V.G. history of Rangoon. Neale had become First Squadron leader after the death of Bob Sandell, who died flight-testing a repaired P-40 over Mingaladon. Since the fall of Rangoon was already looming, Neale no longer retained damaged planes at Mingaladon but had them flown or shipped north by rail. About this time I also ordered Neale to cease all strafing and bomber escort missions due to the worn out condition of the P-40 engines, which were long overdue for overhaul. (Chennault, *Way of a Fighter*, 137)

Figure 9–1. Remains of a British Buffalo. Caught on the ground and burned by a Japanese raid at Mingaladon. (Neale, 225)

AVG and RAF units were finished as an effective fighting force. AVG pilots knew it, RAF pilots knew it and the Japanese knew it too. There was little to do as the Americans say "but get the hell out of Dodge."

> When the Japanese began their final aerial assault on February 26, there were only 15 Allied fighters to meet the attack by 166 enemy planes. They fought off three raids on the twenty-fifth with the A.V.G. bagging 24 Jap planes. The next day was even worse, with 200 enemy planes over Rangoon. The A.V.G., now reduced to 6 P40's, bagged 18 Jap fighters to bring their two-day total to 43 enemy aircraft without loss to themselves. (Chennault, *Way of a Fighter*, 138)

Even God had to rest on the seventh day, and so it was for Bob Neale and his 1st Squadron pilots. They had put up a magnificent fight, one that will probably never be equaled, but now it was time to go.

> On the night of February 27 the RAF removed the radar set from Rangoon without previous notice to the A.V.G. For Neale that was the last straw. The next morning he sent four of his remaining six P-40's to cover the route of the last A.V.G. truck convoy to leave Rangoon. He and his wingman, R.T. Smith, later an A.A.F. fighter group commander, stayed to make a final search for an A.V.G. pilot who had bailed out over the jungle some days before. Neale ripped out his own radio and enlarged the baggage compartment to hold a stretcher case if the pilot turned up injured. Neale and Smith sweated out February 28 waiting for news of the lost pilot, Edward Liebolt. The next day the Japanese cut the Prome Road, the last landline of retreat from Rangoon.

> Neale and Smith jammed two cases of whiskey into Neale's baggage compartment and took off for Magwe, two hundred miles to the north. (Chennault, *Way of a Fighter*, 139)

Magwe was an RAF base in an arid part of central Burma along the Irrawaddy River. There were few facilities, but it was a long way from Rangoon. It took the Japanese a while to find it.

Rangoon anarchy

Rangoon's fate was now determined. Panic, anarchy and wholesale looting began. It started in the dock area. Materials and supplies of all sorts were hauled off in any kind of transport available. Foreigners on the street had to be armed and prepared to fight or die. Most of the foreign community were long gone, except for the unfortunate few left behind because of their work.

Everyone tried to find a way out before the Japanese Army surrounded Rangoon, cutting off the last exit. Prisons opened, armed gangs roamed the town raping and murdering at will. Police fled while they could. All insane asylums and leper hospitals were emptied. Mob mentality took over and Rangoon became a little patch of hell on earth.

At the last moments of civil order, the few remaining military units in Rangoon burned docks and warehouses. These were stuffed to the rafters with critical military supplies. China needed any supplies available and so did the AVG. Chennault ordered his men to take anything they could use, including trucks and jeeps, sending them north to Kunming.

For weeks, every man who was able traveled to Rangoon. They spent their days and nights sweating, working at the docks until the sun came up. Even Chaplain Frillman worked like a Chinese coolie. Every day he gathered whatever ground crewmen he could scrounge and continued loading every vehicle he could steal.

Men were busy smashing open crates, barrels, loading trucks with spare parts for planes and vehicles. Tires, tools, radio equipment, guns, ammunition and all the equipment necessary to conduct combat operations were loaded and sent north. They gathered a three-month supply of canned goods and enough liquor for the whole AVG.

When time ran out, they just snatched whatever crates were handy and portable with no idea of the contents.

This was exciting, just like a little kid told to go into a candy store and take whatever he wanted. The downside of AVG participation in these actions was that some AVG men engaged in wholesale black market operations and other criminal activities. The free-wheeling inducement to "take what you want" tainted the AVG reputation. Many men, Chennault included, were accused of piracy, black marketeering and larceny. This indeed happened in some instances and a few of the AVG men freely participated in these larcenous acts, accruing larger proportions of guilt for the entire group than deserved.

> AVG personnel were flown down from Kunming to drive convoys of trucks north to the Burma Road and then over that winding, traffic-filled stretch to Kunming. Chennault arranged with Chinese customs officials at the border with Burma to pass all AVG trucks and convoys without inspection, having assured them that only supplies being brought into China were for the AVG's use only and not for sale on the black market. (Schultz, *The Maverick War*, 189)

It passed into rumor. Chennault the military man, an officer and a gentleman, a man who did not covet material wealth, was accused of larceny, profiteering and black market operations. Nothing was further from the truth; it was only emergency procurement of combat requirements under extraordinary circumstances.

Chennault authorized pilfering of government stores for military usage only. Those few that did steal brought lasting discredit to themselves. AVG members who carried out actions for personal profit did a disservice to everyone in the group. Material was taken for non-military purposes by some men who were under Chennault's command. This was theft for personal gain and correctly condemned. Enough said about this matter.

Rangoon falls

The Japanese army entered Rangoon unopposed on March 8, 1942. In the eyes of the Japanese High Command, the fall of Rangoon was a major victory, this was the win that would allow the Japanese to conquer Burma. In the Japanese war plan for South-east Asia, Burma was the hinge-pin of victory against both the British and Chinese.

With control of Burma, the Japanese intended to implement phase one of their "Domino Theory," with the fall of Burma it most certainly would follow that both India and China would be defeated. This was the cap-stone in the Japanese path to victory in Asia and it would lead to the defeat of the Allies. With Asia subjugated the Allies would have no choice other than to sue for peace, the Japanese position would be unassailable. So seizing the rest of Burma was paramount.

Allied reorganization

With the United States entering the war, who would command American and Chinese forces, and where would Chennault fit in? Within the military of 1942, there were only two men that had the knowledge and experience to deal with the difficult conditions of the

Figure 9–2. Victorious Japanese in Rangoon. The city fell on March 8, 1942. Government House, the seat of British government in Burma, provides a backdrop for the Japanese celebration of their victory. (Fold3, 5306253)

Chinese government and the existing state of military affairs. On the surface both were eminently qualified to hold the senior military position of theater commander.

On one hand there was Chennault, still considered a renegade and troublemaker by the military establishment. He was not a graduate of West Point, he was not a team player and his abilities to hold this command were questioned. His main qualifications came from his dealing with the Chinese government and his relationship with the leaders of the Chinese military. In this regard he had done a masterful job, working successfully with the Generalissimo and Madame Chiang Kai-shek.

Although facing many difficult military and political crises, Chennault had always found an equitable and satisfactory solution to these problems. He was held in the highest esteem by the Generalissimo and Madame and had their complete trust, they knew that his love of the Chinese people was genuine and heartfelt. He had proved both his genius and loyalty numerous times. Chennault had shown himself to be an open and honest person and had never attempted to profit personally from the volatile situations he faced daily. He held the esteem and loyalty of the Chinese military and was openly loved and admired by the Chinese people. These are qualifications that could not be duplicated by any then serving officer in the United States military. If he was put in charge of military operations it would guarantee a smooth and efficient transition for the American forces and not adversely affect the war effort in China. Chennault was already using an effective war fighting strategy that has been proven effective.

The only other officer that had any qualifications was Major General Joseph Stilwell. He has been deployed in China since the early 1930's and also knew the Chinese military and political leadership. His primary qualifications were that he spoke fluent Chinese and has been a first-hand witness to all the upheavals that had occurred both militarily and politically in China for the last ten years. He had observed and advised the Chinese military in their anti-communist war and the numerous battles against local warlords and the many bandit groups in China.

Unfortunately for the Generalissimo and China, Stilwell had no respect for the Chinese leadership, their military, or the Chinese people in general. Stilwell considered the Chinese government to be corrupt and self-serving, he considered the Chinese military to be incompetent, poorly organized and led and incapable of any kind of military success at all. He found the Chinese people to be lazy, uneducated, dirty and past redemption. Stilwell openly and contemptuously referred to Generalissimo Chiang Kai-shek as "Peanut" because of Chiang's shaved head and he believed that all Madame cared about was money and power.

Stilwell maintained the attitude that was all too common amongst the imperialist Europeans who had come to China to seize as much wealth as they could: they felt China was a country to manage because the Chinese were incapable of doing it themselves. Stilwell was a member of the Old Boys network who had found rapid promotion in Asia, brigadier general in 1939 and major general in 1940, lieutenant general in 1942 and full general in 1944, and a "ring knocker" to boot.

He had many highly placed government and military friends and he could always be counted on to think and act as prescribed by the powers in place in America. At best he was provincial in his outlook with regard to China and at worst he was thoroughly contemptuous of the entire country.

> It was unfortunate that Joe Stilwell was the highest representative the United States had in China. It was unfortunate for the United States and for China. For a great infantry soldier to be plunged into one of the most difficult positions from either a military or diplomatic point of view was a tragedy. (Samson, *Flying Tiger*, 203)

In the jargon of today, you could say that the United States government and the military made the conservative, less provocative and safe choice. Newly promoted Major General Joseph Stilwell (his nickname was "Vinegar Joe"—given with no affection) would be placed in overall command of the American forces in the China-Burma-India (CBI) theater.

Burma retreat

Stilwell arrived in India in mid February and conferred with British authorities, noting the confusing command structure in Burma.

Figure 9–3. Stilwell, Generalissimo and Madame Chiang. Initially, Chiang Kai-shek placed Stilwell in charge of all troops in Burma. The three were all smiles for a photograph in Maymyo, Burma, on April 7, 1942. But as the military situation in Burma deteriorated, so did the relationship between the Generalissimo and Stilwell. (US Army, 134627)

Until now, the British had steadfastly refused Chinese offers of troops to fight in Burma. On March 3, he continued on to Lashio in Burma, meeting Chiang Kiai-Shek and staff there, then on through Kunming to Chungking. After a week getting staff and responsibilities nominally assigned, he returned to Burma and began to formulate a plan for defense.

> In March, therefore, the British combined planes and forces with the Chinese in an attempt to contain the Japanese in southern Burma. Under the command of Lt. Gen. Sir Harold R.L.G. Alexander, the British and Indian troops established two defense lines, one west and another north of Rangoon. The Generalissimo deployed Chinese troops to positions east of the city and placed them under command of Lt. Gen. Joseph W. Stilwell. (Bond and Anderson, *A Flying Tiger's Diary*, 117)

Chengtu
Sadiya
Ledo
Kohima
Imphal
Myitkyina
Yangtze River
Yunnan-Yi
Kunming
Paoshan
Loiwing
Lashio
Mengtze
Mekong River
Salween River
Mandalay
Maymyo
Namsang
B U R M A
Magwe
Sittang River
Loikaw
F R E N
I N D
C H I N
Akyab
Kyedaw
Irrawaddy River
Toungoo
Chiang Mai
Mekong
Pegu
Lampang
Mingaladon
Salween River
T H A I L A N D
Tak
Phitsanulok
Rangoon
Moulmein
Tavoy
Bangkok

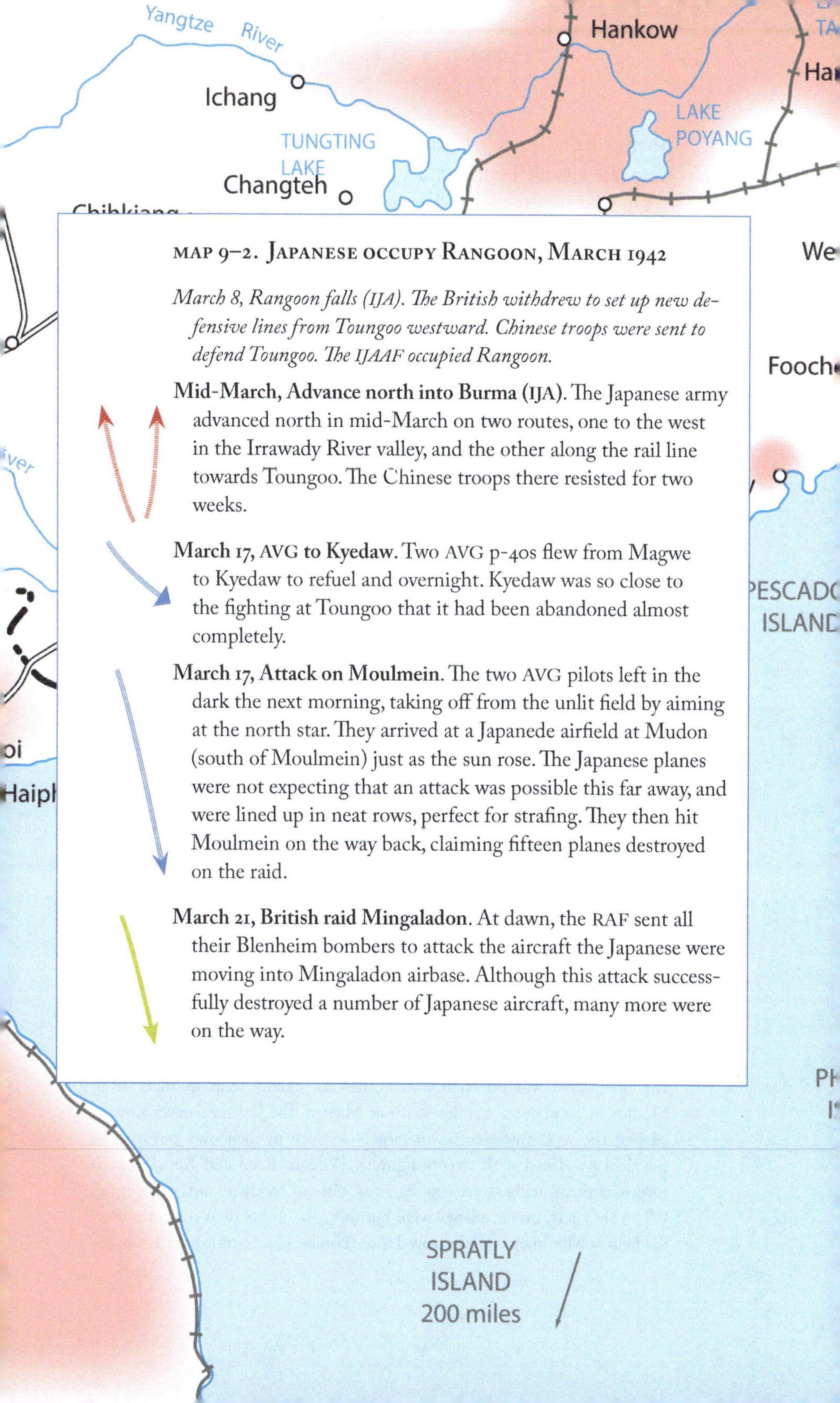

MAP 9–2. JAPANESE OCCUPY RANGOON, MARCH 1942

March 8, Rangoon falls (IJA). The British withdrew to set up new defensive lines from Toungoo westward. Chinese troops were sent to defend Toungoo. The IJAAF occupied Rangoon.

Mid-March, Advance north into Burma (IJA). The Japanese army advanced north in mid-March on two routes, one to the west in the Irrawady River valley, and the other along the rail line towards Toungoo. The Chinese troops there resisted for two weeks.

March 17, AVG to Kyedaw. Two AVG p-40s flew from Magwe to Kyedaw to refuel and overnight. Kyedaw was so close to the fighting at Toungoo that it had been abandoned almost completely.

March 17, Attack on Moulmein. The two AVG pilots left in the dark the next morning, taking off from the unlit field by aiming at the north star. They arrived at a Japanede airfield at Mudon (south of Moulmein) just as the sun rose. The Japanese planes were not expecting that an attack was possible this far away, and were lined up in neat rows, perfect for strafing. They then hit Moulmein on the way back, claiming fifteen planes destroyed on the raid.

March 21, British raid Mingaladon. At dawn, the RAF sent all their Blenheim bombers to attack the aircraft the Japanese were moving into Mingaladon airbase. Although this attack successfully destroyed a number of Japanese aircraft, many more were on the way.

So the British were to defend the area around Prome, and the Chinese the city of Toungoo. By mid-March, Stilwell returned briefly to Chungking to ask Chiang Kai-shek for more troops. Back in Burma, he found that his orders were frequently disregarded or delayed by the various Chinese commanders. He also moved his headquarters further west from Lashio to the British hill-town of Maymyo, the summertime seat of government. Some there hoped that the monsoon rains, which usually started in May and created a sea of mud on the roads, would prevent any serious Japanese advance this year. Stilwell returned to Chungking March 31 to get help with his command issues. Chiang Kai-shek not only placed him unequivocally in charge of the Chinese armies, but also traveled to Maymyo in Burma personally to deliver that news.

The AVG at Magwe

After Rangoon fell, the air battle continued over Northern Burma from the base in Magwe. Japan had 400–500 planes against 30 serviceable AVG fighters and a dozen outdated RAF Blenheim bombers. There was a brief lull in early March while the Japanese occupied the former RAF base at Mingaladon, and repositioned army units for a drive north into central Burma.

> Faced with such odds and realizing that one battle could wipe out a large portion of his planes, Chennault shifted tactics during March. He ordered raids on Japanese airfields, hoping to surprise the enemy and destroy planes on the ground. (Bond and Anderson, *A Flying Tiger's Diary*, 117)

Chennault instinctively understood his enemy and anticipated a major attack was coming. Spies and his warning net let him know where the Japanese forces were concentrated. Chennault implemented bold strikes against Japanese Air Force units inside Burma and Thailand.

> The first strike was on March 19 against an enemy strip 10 miles from Moulmein, or about 250 miles south of Magwe. The Japanese never suspected that the AVG was able to hit targets so deep in their own territory, so the field was lined with twenty fighters. William Reed and Ken Jernstedt surprised them, making six strafing runs without receiving anti-aircraft fire. When they left, fifteen planes were burning. The pilots then turned toward the field at Moulmein and destroyed three bombers and a transport. In all, the

Figure 9–4. Major General Joseph "Vinegar Joe" Stilwell. Stilwell spent much of his early career in China in the 1930s and spoke Chinese fluently. He seemed an ideal candidate to lead the Allied effort in China and Burma during WWII. But his disdain for Chiang Kai-shek and Chinese customs hampered him. He felt that the Nationalist government and culture was rife with corruption. Stilwell believed that the Chiang government remained focused on defeating the Communists, therefore limiting their commitment to fighting the Japanese. (US Army, from Wikipedia)

> raid destroyed nineteen planes, meaning that Reed and Jernstedt recorded the highest number of kills on one mission of any pilots in the AVG. (Bond and Anderson, *A Flying Tiger's Diary*, 117–118)

Flown by only two pilots, this was a tremendous accomplishment for Chennault and the AVG. Looked at from a different perspective, it was a defeat suffered by the Japanese, but there was a reason so many Japanese aircraft were there.

RAF raids Mingaladon

An RAF observation plane discovered a large concentration of Japanese aircraft assembled at Mingaladon, preparing for a major air raid. All serviceable RAF aircraft were readied.

> The next day the British ordered a raid. RAF bombers struck their former base, the aerodrome at Mingaladon. They surprised the enemy, who left fifty planes exposed on the field. Some Japanese fighters got off the ground and damaged a few Blenheim bombers, but nevertheless the British destroyed a dozen enemy fighters and sixteen bombers. (Bond and Anderson, *A Flying Tiger's Diary*, 118)

A substantial victory for the British, their mission planning was flawless. Brilliantly flown by RAF pilots, the mission did significant damage. The Blenheim bomber was a very old airplane designed in the 1920s and already obsolete. They were never meant for service in the Second World War, especially against such first-rate pilots as the 1942 Japanese. Unknown to the RAF and AVG, these losses only dented the armada the Japanese air force was assembling.

Japanese army advances north from Rangoon

In mid-March, the Japanese army attacked central Burma along the two rail lines that headed north from Rangoon. The western one followed the Irrawaddy valley to the oil fields near Prome that were defended by British forces. The other followed the main rail line north towards Mandalay, which was defended by Chinese forces positioned near Toungoo. This battle started March 19, but the Chinese army fought fiercely and held the city for two weeks.

No one expected the entire situation to become a disaster shortly, but it did. As the Japanese pressed north, British civilians and the

Indians they employed faced not only the wrath of the long-subjugated Burmese, but that of the new foe.

> Fear spread after the battle, fanned by the same savagery the enemy had practiced in China. Indian prisoners were tied by the Japanese in bamboo houses and set afire or soaked in gasoline and burned alive. More shocking and frightening to Westerners in Burma was the fact that captured British officers were used no better; in some cases, they were stripped and tied to trees for bayonet practice. Until now, every Westerner who ever entered Asia had taken it for granted that he would receive different treatment than a native. (Tuchman, *Stilwell the American Experience*, 284)

10

And Then There Was Hell to Pay

Any hope that the Japanese, having severed the last supply route to China, would wait out the summer monsoon before advancing disappeared in late March.

> Alexander, Stilwell, Chennault and Chiang Kai-shek faced a powerful Japanese military machine in Southeast Asia. The Japanese army was larger, better equipped, and reinforced with ample air support. Fourteen enemy air regiments were based in Thailand and southern Burma, meaning that about 450 enemy planes confronted a dozen RAF bombers and thirty AVG fighters, which were based at Magwe in central Burma. (Bond and Anderson, *A Flying Tiger's Diary*, 117)

The destruction of Magwe air base

The Japanese air force was preparing to strike a major blow to the RAF and AVG at Magwe. They discovered the location of the remaining RAF and AVG forces on March 9, but chose to assemble all their aircraft and deal a decisive attack. The two RAF and AVG air raids were puny by comparison.

> The Japanese responded swiftly and forcefully. On March 21, they initiated a massive attack against Magwe. For twenty-five hours, 266 enemy planes pounded the airfield. Two waves of bombers struck first, destroying many Hurricanes and Blenheim bombers. A few hours later they attacked again, leaving the field smoldering and fatally wounding two members of the AVG,

> pilot Frank Swartz and crew chief John Fauth. The attack on Magwe eliminated the bombing capability of the RAF in Southeast Asia. (Bond and Anderson, *A Flying Tiger's Diary*, 118)

Because the Japanese fighters and bombers came from so many locations and arrived in waves, the field was under continuous attack for several hours and there was no chance to refuel and rearm. After the first wave, the field was completely undefended.

The next day, the Japanese repeated this raid, flying the same routes. Due to the previous day's damage, the RAF and AVG got no warning, and no fighters defended Magwe. By the next morning, only four Tomahawks were fit to fly, and the AVG retreated to Pawley's factory at Loiwing.

AVG raid on Chiang Mai

This Japanese reprisal raid was unanticipated and had a significant effect on Chennault and his men.

> Chennault and his men were enraged, and he called Bob Neale and Jack Newkirk into his office to discuss plans for retaliation. The Old Man knew the headquarters of the Japanese Air Force in Southeast Asia was at Chiang Mai, Thailand, and he selected it as the primary target with a secondary objective a smaller base some twenty miles southeast of Lampang. An overnight stop at the RAF airbase of Namsang, Burma, would permit a predawn take-off the morning of the strike. (Bond and Anderson, *A Flying Tiger's Diary*, 118)

Charley Bond, in his diary on March 24, 1942, described the day that followed:

> Bartling, who had never flown an aircraft at night, was to join up with Boyington. As he approached ten thousand feet and cuddled up next to his flight leader, he saw the number 7, Bob Neale's P-40. He dropped back and I slipped into place. Pappy was close behind, and soon we were in formation making a last circle around the field just as dawn was beginning to break across the eastern mountain tops. (Bond and Anderson, *A Flying Tiger's Diary*, 133)

Here is a perfect example of how little training some of Chennault's flyers had. Mission requirements forced inexperienced or untrained pilots to fly some of the most dangerous missions possible. Chennault did not have the luxury of deploying trained, competent pilots. In his war, he had to fight with whatever resources were at hand.

> The early morning haze hid any landmark that would have been helpful, but I could not miss the deep Salween River gorge as we passed over it. I estimated a ground speed of 230 M.P.H. and figured Chiang Mai Airfield at 7:12. (Bond and Anderson, *A Flying Tiger's Diary*, 133)

See how the Second World War began for America? Flying inferior aircraft at night with pilots who had never flown formation in the dark was a desperate action. You can believe me, night formation is very difficult when you are untrained—and very dangerous. Not many atheists fly these types of missions. Lots of praying in the dark. This attack illustrates the courage if not the skill level of Chennault's Tigers. At night, over unfamiliar mountainous terrain without navigation aids or maps requires suicidal courage. These are the types of missions that keep young pilots from sleeping. Long odds require great risks. One might well ask from where do we get such men?

> I steadied my aim and strafed the entire line of ships. I was so low that it seemed to me as though my prop tips would clip the heads of Jap pilots who by now were trying to get into their cockpits. Other men, evidently maintenance people, were laying flat on the ground nearby. Props were turning on some of the fighters, but not a single aircraft was taxiing. As I completed my second run, I looked at the area I had strafed first: three aircraft were in flames. (Bond and Anderson, *A Flying Tiger's Diary*, 134)

Combat is a totally consuming occupation. All concepts of time, location or attitude of the airplane are gone. The pilot sees only what is directly in-front of his gunsight. Charley's diary entry continues:

> Only seven or eight minutes had passed, and I had seen only one P-40, Bartling's. Banking again, I caught a glimpse of three aircraft to my right. (Bond and Anderson, *A Flying Tiger's Diary*, 134)

Destruction from successive passes in that ten to twelve minutes resembled hell on earth. Burning aircraft looked like the entry to Hades.

> The entire airfield seemed to be in flames as we headed back to Burma. We apparently caught them as they were preparing a major concentration of aircraft for a big strike against us in China. It was a mission accomplished exactly ac-

Chengtu
Yangtze River
Sadiya
Ledo
Kohima
Myitkyina
Imphal
Yunnan-Yi
Cha
Kunming
Paoshan
Loiwing
Mekong River
Lashio
Mengtze
Mandalay
Maymyo
Salween River
Namsang
BURMA
Magwe
Sittang River
Loikaw
FREN
IND
CHIN
Akyab
Kyedaw
Irrawaddy River
Toungoo
Chiang Mai
Mekong
Salween River
Pegu
Lampang
Mingaladon
THAILAND
Rangoon
Tak
Phitsanulok
Moulmein
Tavoy
Bangkok

MAP 10–1. MAGWE WIPED OUT, LATE MARCH 1942

March 21 & 22. Despite the morning RAF raid on Mingaladon, in the afternoon of the 21st the IJAAF attacked the airbase at Magwe with waves of fighters and bombers coming from bases in Thailand and Burma. The sheer number of Japanese aircraft overwhelmed the RAF and AVG and the field was devastated. The next day, they attacked again. Only four P-40s could be salvaged.

March 24. Chennault planned a retalation raid on Chiang Mai, a forward airfield for the Japanese. On March 23, 10 P-40s left Kunming, paused at Loiwing, then headed for the British airfield at Namsang, Burma. Because gas was stockpiled there, the Tigers could refuel and take off again before dawn. They caught the Japanese still on the field.

Japanese Army makes steady progress. The IJA continued attacking Stilwell's defensive lines, which gave way at the end of March.

cording to Chennault's plan. How proud he would be. Hell I was! (Bond and Anderson, *A Flying Tiger's Diary*, 135)

Audacity, always with audacity. These words should be emblazoned on Chennault's tombstone. Look up the definition of audacity in the dictionary and you will see a picture of Chiang Mai! It took genius to conceive and courage to carry out. No one in Asia except Chennault would attempt it. Even considering the hype factor, it had an indelible effect on the morale of the American public. When no one else successfully engaged the Japanese, there was one small band of American airmen and their support crews taking it to the vaunted Japanese war machine.

It is difficult to see how Chennault or the men of the AVG could do more. In a time of great darkness, Chennault and his men were America's only light.

We made a screaming dive and buzzed the field in a victory formation pass. Then in a sharp climbing turn to string out for landing. (Bond and Anderson, *A Flying Tiger's Diary*, 136)

Imagine the sense of relief these men felt. Flying a very dangerous mission and returning home alive. The release of tension and fear is unmatched in human experience. Now was the time for counting, as they say, now that the fighting was done.

> Once on the ground, we got together and assembled the AVGers at Loiwing to recap our experience and critique the mission. We found good additional news–and also additional bad news. Jack Newkirk's flight of four had rendezvoused after takeoff at about six thousand feet over the field and then headed out. This put them in the Chiang Mai area a few minutes ahead of us. (Bond and Anderson, *A Flying Tiger's Diary*, 136–137)

Death comes stalking

But strafing was dangerous business, being in range of antiairdraft guns and even rifle fire. Here is claimed the squadron leader.

> Jack Newkirk, according to Geselbracht and Lawlor, had apparently spotted what appeared to be a Japanese tank or armored vehicle and went down to strafe it. He never pulled out. (A Flying Tiger's Diary, *p. 137)*

"Scarsdale" Jack Newkirk died in a blaze of glory as his stunned flight watched. The news deeply saddened Chennault and his men, but they soldiered on. Death in combat comes quickly and with awesome finality.

There are several noteworthy items revealed in this passage. Combat is a very fluid thing and can change in an instant. Here, a squadron leader flew a secondary mission when the primary was unavailable. He chose aggressive action against a back-up target. A dangerous low level strafing mission against already alerted ground forces. Being low to the ground, he exposed himself and his flight to increased counter fire in an extremely hostile area. The difficulty of command weighs heavily on the commander.

Another thing seen in this attack is a problem all tactical pilots face. It is a killer known as target fixation. When a pilot becomes locked onto his target, he loses situational awareness and sometimes flies into the ground. Always a present factor, in this case it probably killed Newkirk. A flight leader in combat needs aggression, flexibility and willingness to make critical decisions instantly. In all combat operations, danger is present and the smallest mistake might cost you or your flight members their lives. Flying is a dangerous occupation, combat flying even more so. These men were the bravest of the brave.

> Casualties were mounting. Frank Swartz and John Fauth died after the Japanese raid on Magwe, and William McGarry wandered in the jungle for twenty-eight days before he met another human–Thai officials who arrested him and put him in jail in Bangkok. (Bond and Anderson, *A Flying Tiger's Diary*, 137)

In terms of enemy aircraft destroyed, the mission is rightly called a victory. Victory came with an extravagant cost. These men were part of the heart and soul of the Group. They were sorely missed. Scarsdale Jack Newkirk and Black Mac McGarry's leadership, personalities, humor and spirit lingered like ghosts in the night.

Disaster on the ground

The two Japanese armies in the center and west of Burma continued to advance, but more slowly against British and Chinese resistance. At this point in the central Burma campaign, the AVG and RAF had too few planes to significantly slow the advancing Japanese ground forces. The Japanese maintained air superiority where their armies were fighting, and continued attacking the Chinese at Toungoo. The Chinese held out until March 30, but were forced to withdraw north from Toungoo along the Sittang River.

The British at Prome fared worse when the Japanese Army attacked March 30. They withdrew north on April 2 towards Magwe and the oil fields of Yenangyaung nearby. Stilwell drew up new lines of defense for the British at Magwe, and for the Chinese near Pyinmana.

> On the Irrawaddy front three days later [April 14], the Japanese broke through, bypassing the 1st Burma division and heading for the oilfields at Yenangyaung. General Slim could get no effective action out of the demoralized troops summoned by Alexander to Maymyo, found "disaster and gloom." Alexander confessed to him that his men were "simply afraid of the Japs." (Tuchman, *Stilwell the American Experience*, 284)

The advancing Japanese encircled part of the British army in the oil fields at Yenangyaung. Stilwell arranged for a particularly able

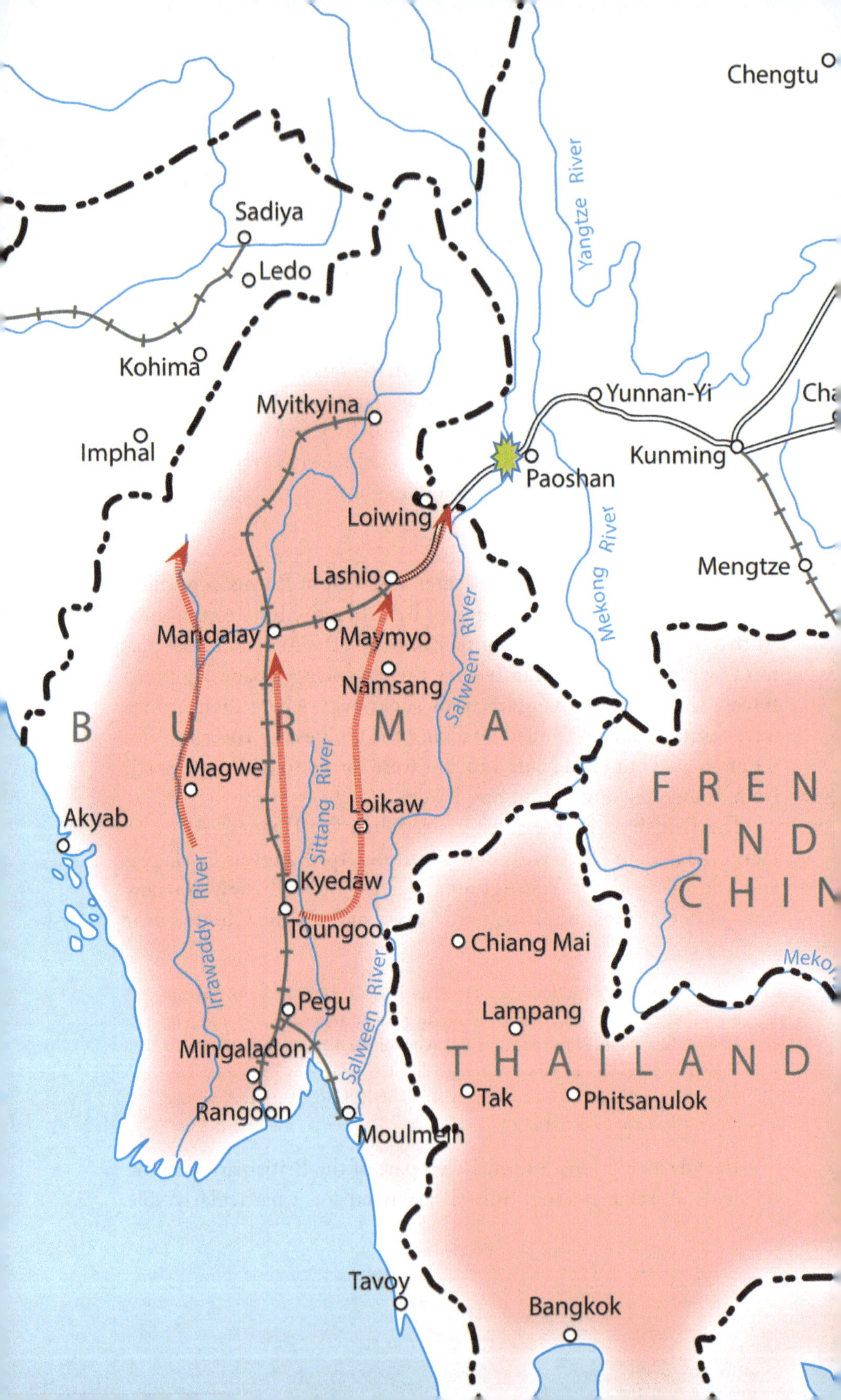
Chengtu
Yangtze River
Sadiya
Ledo
Kohima
Myitkyina
Imphal
Yunnan-Yi
Kunming
Paoshan
Loiwing
Lashio
Mengtze
Mekong River
Mandalay
Maymyo
Namsang
Salween River
B U R M A
Magwe
Sittang River
Loikaw
Akyab
Irrawaddy River
Kyedaw
Toungoo
Chiang Mai
Pegu
Salween River
Lampang
Mingaladon
T H A I L A N D
Tak
Phitsanulok
Rangoon
Moulmein
Tavoy
Bangkok

MAP 10–2. RETREAT FROM BURMA, APRIL & MAY 1942

April 20. Stilwell learns that a Japanese army is in the eastern valley moving north of Loikaw, threatening to outflank his defensive lines. Efforts over the next week to stop them fail, and Stilwell's armies begin to retreat northward.

April 29. The flanking Japanese army captures Lashio, astride the Burma Road. This blocks any possibility of retreat to China.

April 30. The center Japanese army approaches Mandalay as the retreating British and Chinese disintegrate into a rout.

May. All Japanese armies continued northward, taking control of most of Burma. Some of the Japanese army advancing up the Irrawaddy River continued up the Chindwin River to interrupt the retreat, while another part took Akyab.

In May, fleeing soldiers and civilians walked through jungles and over the mountains into India using whatever paths are available. Some Chinese armies retreated to China through the mountains of north Burma.

May 4. A Japanese regiment continued along the Burma Road towards the last defensible feature, the Salween River crossing. Their advance was slowed because the Burma Road was choked with people trying to walk to China.

May 5. The Salween River Bridge was destroyed in hopes of stopping the Japanese advance.

Chinese unit save them. The Chinese did so successfully. However, the real problem had not yet surfaced.

The surrender of Singapore had allowed the Japanese to move army forces from Malaya to Burma. A mobile mechanized and armored unit arrived at Toungoo just after its capture. On April 4, this crack unit moved out of Toungoo and crossed the intact bridge over the Sittang River. They continued on good roads through the mountains to the east of the Sittang River and then turned north. The Chinese force defending the valley leading north was under strength, poorly equipped and poorly led. The Japanese tanks followed good roads while meeting light, unsuccessful resistance from Chinese troops. Most importantly, its advance went unreported to Stilwell until April 20. It rapidly outflanked his defensive lines and threatened the Burma Road at Lashio.

Stilwell's efforts to halt the eastern advance proved unsuccessful, and all his armies started retreating. When the eastern Japanese army captured Lashio on April 29, the retreat became a rout northward. With no hope of using the Burma Road to reach the safety of China, the only other choice was to walk out to India. There were no roads, only paths leading through jungles, across unbridged rivers and over mountains. Many perished on this arduous trek.

Threat to China

There was no effective Chinese army to stop an advance along the Burma Road. Wasting no time, the Japanese sent a regiment along the road east from Lashio in early May. The masses of people fleeing Burma slowed their advance. The Salween River Bridge was destroyed on May 5, hoping to stop the Japanese advance. But if they crossed that river, nothing stood in the way of the Japanese continuing on to Kunming. It would be the downfall of China.

Once China fell, Japan would begin attacks on the Indian subcontinent. With the defeat of India, the total balance of world-power shifted to Japan and her partners. Isolated and cut off, the Allies and China must surrender or face total destruction. Japan was on the verge of victory in Asia.

11

On the Road to Victory

With control of Burma, the Japanese could launch attacks into the interior of China via the Burma Road and down to the Salween Bridge. Once past this last strategic barrier, China's new wartime capital Chungking was in peril. They thought the Chinese government must sue for peace. With China subdued, Japan could maintain minimum garrison troops throughout the country and deploy the rest of their forces (now including millions of Chinese levies) in support of other planned offensives.

Freedom to rampage deep into the Indian subcontinent or the Pacific were Japan's other options. Possession of the Solomon Islands and Port Moresby allowed the possibility of invading Australia. Control of India split the British Empire and provided a limitless supply of men and material for the Japanese war machine. Here was the capstone of Japanese War Plans. With these goals attained, Japan's position in the Pacific and Asia was unassailable. The Allies must surely sue for peace.

The unexpected

Before the Japanese could launch their ambitious offensive, however, two items of extreme significance came into play. The results accruing from these unforeseen actions changed the course of the war in the Pacific and China, causing the death of additional millions of innocent Chinese.

After the "unprovoked and dastardly attack by Japan on Sunday, December 7, 1941," President Roosevelt sought a means of quickly getting at the Japanese. He issued an order to Admiral Ernest King, Chief of Naval Operations, to pursue any action allowing forces of the United States to come to grips with the Japanese.

Admiral King put "Bull" Halsey in charge of the Pacific fleet. Halsey was aggressive, and willing to use his two newest and fastest carriers (USS *Hornet* and *Enterprise*) to strike Japanese targets. Halsey oversaw several unconventional operations. Limited attacks against small island garrisons like Makin Atoll caught the Japanese by surprise but had little impact. But he succeeded in training his command to strike fast and get away.

Doolittle Raid

The most significant action involved a daring raid planned between the Navy and the United States Army Air Corps. Sixteen B-25 medium land bombers would launch from the flight deck of the *Hornet.* The newest and largest U.S. aircraft carrier was just big enough to accommodate the strike force. America's greatest pre-war airman, Lt. Colonel Jimmy Doolittle, commanded the American force bombing Japan.

With hand-picked, specially trained airmen, Doolittle mounted his B-25 bombers on the flight deck of the aircraft carrier *Hornet* at the Oakland Navy Yard. From there, the task force would sail to a predesignated launch point 600 miles off the coast of Japan.

Doolittle's aircraft targeted six specified Japanese cities. Pilots were ordered to bomb strategic war production facilities or military targets only. Mission accomplished, the bombers were to continue on to airbases in China. These planes were destined for Chennault's use in follow-on bombing missions against the Japanese Home Islands. At least, that was the plan.

Unfortunately, Japanese picket boats spotted Halsey's force early. He now had to make a critical decision in a hurry. Doolittle and his raiders faced difficult circumstances. The whole mission was in jeopardy. Halsey, with Doolittle's consent, made the tough choice. If

FIGURE 11–1. B-25 LAUNCHING FROM *HORNET*. The Doolittle Raid was nearly impossible. The planes barely fit on the flight deck, and the runaway was laughably short. The pilots received extensive training to learn to accelerate and take off in such limited space, and they also needed to launch into a strong wind while the ship moved at top speed. All 16 bombers launched successfully. (NARA, 80-G-41196)

FIGURE 11–2. VIEW OF *HORNET* FROM *ENTERPRISE*. A B-25 piloted by a Doolittle Raider heads for Tokyo. The second ship was necessary to provide combat air patrol and fighter escort. (NARA, 80-G-41197)

immediately launched, the B-25s could bomb designated targets but would not have sufficient fuel remaining to reach the Chinese mainland. Doolittle faced the loss of his entire force.

If the fleet continued to the pre-designated launch point, the possibility existed that the entire striking force might be discovered and destroyed. Without hesitation, Halsey and Doolittle ordered, "Pilots, man your planes." On April 18, 1942, Doolittle launched what he considered a suicide mission.

As originally planned, the attack would have reached a landfall at targets in Japan in the late afternoon. Gathering darkness would prevent interception, allowing the safety of a night flight to China and an early morning landing at an airfield behind Japanese lines in China called Sui-chan. The forced early launch now required Doolittle's aircraft to begin the attack at midday Tokyo time.

Increased distances now allowed just enough fuel to reach China, maybe. Arriving in the dark, there was no way Doolittle's aircraft could land at night over unfamiliar territory on remote airfields. Displaying incredible determination and grit, all sixteen B-25s dropped ordinance on designated Japanese targets then proceeded en route to the Chinese coast. Several aircraft ran out of gas and ditched in the ocean just off the mainland. The remaining crews parachuted over land or crashed in the night. One aircraft flew to Russian territory and landed safely. As a non-belligerent at that time, Russia interned that crew for many months. Later, the American aviators returned to the United States.

> Back home, news of the Doolittle Raid electrified the public. After taking it on the chin for four months, America had landed a blow of its own. Headlines screamed, American planes bomb Tokyo... Even Henry Stimson, a skeptic of the plan, admitted that the raid produced a very good psychological effect in the country, both here and abroad. (Jordan, *American Warlords*, 157)

> The Doolittle Raid was the kind of dramatic statement Roosevelt loved to make, at least when there was nothing else for him to do. When asked by reporters from where the bombers had launched, he gave them an inscrutable grin. Referring to the mythical realm in Hilton's *Lost Horizon*, he replied, "they came from our new secret base at Shangri-La!" (Jordan, *American Warlords*, 157)

As far as Chennault knew, this might be a truthful answer. Fearing for the integrity of the American code breakers' success on the JN-25 code, the Navy and Air Corps did not reveal to Chennault or Chiang Kai-shek that the raid was coming. Chennault raged at not being told about the attack. He fumed over the loss of the sixteen B-25 bombers that he was promised. Chennault claimed had he known in advance of the attack his Chinese warning net could have vectored the lost B-25s to safety (they probably could have).

God Bless the Chinese peasants

Doolittle's aircrews, stranded in small groups, were lost in the expanse of China with enemy troops all around them. These men experienced a period of extreme terror. Airmen caught by the Japanese expected no mercy and possibly a quick execution at the end of a sword. This was indeed the fate of several captured Raiders.

For most of Doolittle's airmen, a kinder fate awaited. Recognized by the Chinese as friends in combat, these men were fed, protected, and given medical treatment for their injuries. An underground network sprang up around them.

Raiders were hustled from place to place, often just ahead of the pursuing Japanese. American flyers were always aided by the Chinese people. Peasants knew full well what Japanese soldiers would do to them if they helped Americans escape. They helped anyway. Now began one of the most gallant acts of the war in China. These unheralded people gave their all to aid Doolittle's downed airmen. Later, they faced horrific consequences from the Japanese Army for their defiance.

Consequences of the Doolittle Raid

May 1942 started the time known in China as Japan's "Three All" Policy (*sanko sakusen*). Conducted by brave, noble Japanese warriors, it was "Steal all, Burn all, Kill all." These rampaging Japanese inflicted extermination on helpless Chinese civilians. Raping and pillaging everywhere they went, killing every living creature found. In affected villages, towns and cities, Japanese soldiers looted and burned every structure, leaving no creature alive. Imperial Headquarter Order number 575 sanctioned these types of action on December 3, 1941.

Tokyo knew the type of war to be fought in China and Asia, even if no one else did.

> The Doolittle Raid is another good example of the singularly one-sided view of the Chinese war effort that affected many Americans in China. As a result of the Doolittle Raid, the Japanese sent an expeditionary force of a hundred thousand men, supported by a sizeable air force, to seize the airfields that were to have been used by the Doolittle mission and Force Aquila, a group of B-17s and B-24s that were scheduled to bomb Japan from East China airfields shortly after the Doolittle raid. (Samson, *The Flying Tiger*, 176)

Scorched earth with a vengeance exceeded even the worst orders from irate officers. These sons of Nippon were allowed to murder as they wished. Japanese soldiers would not be punished for any action committed. The order of the day was kill, kill, kill, whether by rifle, bayonet or sword. Gleefully following orders, the Japanese went about their business. By the time this directed attack finished, over 250,000 Chinese civilians were dead. This was the largest slaughter of civilians in China since the Rape of Nanking when that city fell in 1938. Japanese troops ran unrestrictedly amongst the poor Chinese. Streets ran red with blood for days. This was a senseless murder perpetrated with industrial efficiency on defenseless Chinese civilians.

> In a three month campaign, the Japanese drove their bloody spear two hundred miles through the heart of China, devastating 20,000 square miles, plowing up landing fields, and exterminating everybody remotely suspected of aiding the Doolittle raiders. Entire villages through which the raiders passed were slaughtered to the last child and burned to the ground. (Samson, *The Flying Tiger*, 176)

When Doolittle reunited with his survivors, he was despondent, feeling himself a failure. He lost his entire squadron of aircraft while ten percent of his men were missing, captured, or dead. Certain he would be court-martialed and forced out of the military, he steeled himself for the future. Arriving in the States, Doolittle learned the Raiders and their commander were heroes. All received the Distinguished Flying Cross and other medals. Each man received a one rank promotion and returned to their original squadrons. Doolittle was given the rank of Brigadier General and received the Congressional Medal of Honor from President Roosevelt himself.

A less glamorous and far worse welcome awaited the eight raiders captured. These men faced imprisonment and possible death in Japan. Three of them were executed and one died from illness. Perhaps the Doolittle Raid needs to be renamed the Do-Much Raid. It changed military history and the course of the war in the Pacific. Americans felt vindicated and the dangerous undertaking provided an emotional uplift for all Americans. From this point forward, the Pacific war shifted in favor of the Allies. Doolittle and his raiders were the stuff that makes heroes. Well done to all. Never forget what the Chinese gave so Doolittle's Raiders could come home.

Many war-altering actions went forward from this time, actions not anticipated by Japanese militarists. Done without calm consideration for the end consequences, these "ad hoc" missions would later haunt the Japanese. The most significant over-reaction was to expand Japan's island defensive ring further into the Pacific Ocean. This decision necessitated the Japanese conquest of the Solomon Islands (with Guadalcanal) and Midway Island.

In one of his few errors of the war, Chennault considered the Doolittle raid a stunt of little tactical or strategic importance undertaken to placate the sagging American spirit. Nothing was further from the truth, as history revealed after the battle of Midway June 5, 1942. No one realized the Japanese had reached their highwater mark at Midway. Their losses at Midway and Guadalcanal drained resources from Burma and the rest of Southeast Asia that might have been used against China and the AVG.

Solomon Islands

The Solomon Islands lie north-east of Australia, and could provide airfields to threaten Allied shipping to Australia. The Japanese seized Tulagi on May 3–4, 1942, to set up a forward seaplane base. Then in July 1942, they started to construct an airfield on the larger island of Guadalcanal. The Allies surprised the Japanese by invading and seizing the airfield on August 7. Battles for this island became viewed as decisive for both sides, and consumed an ever-increasing proportion of Japanese effort over the next four months.

Battle of Midway

Midway was then a remote base for American scouting and attack aircraft with a small naval servicing facility and a contingent of Marines. The Imperial Japanese Navy prepared a massive fleet to attack, capture and garrison troops on the atoll. Start of this operation was scheduled for early June 1942.

Winds of war are fickle, and here they would now blow fickle for Japan. The Japanese High Command did not know Americans broke their secret naval code (JN-25) and were reading Japanese mail as quickly as the fleet at sea. American forces positioned themselves to intercept an unsuspecting Japanese fleet.

What followed in the confusion of battle can only be described as the luckiest five minutes in American naval history. During this major and confusing two-day action on June 4–5, 1942, all four participating Japanese aircraft carriers (Pearl Harbor attack veterans) were sunk. Along with the carriers, Japan lost the bulk of their highly skilled and irreplaceable Naval aviators. This was a striking loss for the Japanese Navy from which they never fully recovered.

The Battle of Midway was the seminal defeat for Japan in the Pacific. The balance of power shifted to the Allies. From this point forward, Japan never effectively replaced the men or equipment lost in this action. Deprived of critical resources, Japan's military improvised all future actions in the Pacific, Asia and mainland China. Everything Japan's military did in Asia was now unscripted. Stunned by defeat at Midway, the Japanese hurriedly planned future actions and inappropriately responded to perceived Allied threats. Working with inaccurate intelligence and less capable airmen, Japan let the advantage shift to the Allies.

12

Salween Action

With control of Burma, the Japanese could launch attacks into the interior of China via the Burma Road and down to the Salween Bridge. Once past this last strategic barrier, China's new wartime capital Chungking was in peril. They thought the Chinese government must sue for peace. With China subdued, Japan could maintain minimum garrison troops throughout the country and deploy the rest of their forces (now including millions of Chinese levies) in support of other planned offensives.

Freedom to rampage deep into the Indian subcontinent or the Pacific were Japan's other options. Possession of the Solomon Islands and Port Moresby allowed the possibility of invading Australia. Control of India split the British Empire and provided a limitless supply of men and material for the Japanese war machine. Here was the capstone of Japanese War Plans. Once these goals were attained, Japan's position in the Pacific and Asia was unassailable. The Allies must surely sue for peace.

In the spring of 1942, the Japanese had placed a blockade around China that was as effective as a new Great Wall. The seaports and coastline belonged to Japan and all the great cities along the sea were in the hands of the Japanese. Shanghai in the north to Foochow and Amoy in Fukien Province and Swatow and Hong Kong in

FIGURE 12-1. NEW P-40 MODEL E. Improved P-40 Model E fighters began arriving in China in April 1942. This photos shows a new batch lined up on the field at Kunming, October 1942. Featuring six 50-caliber, wing-mounted machine guns and a larger engine, these planes were a strong improvement for strafing over the P-40 Model Bs. To extend their range, each was fitted with a disposable gas tank under each fuselage. The tanks could be dropped once the fuel was used. The Tiger mechanics immediately saw an additional use for the dropping mechanism. They cleverly adapted it to hold a 550-pound Soviet bomb, which could then be dropped on targets during dive bombing runs. These large bombs were used with great success at the Salween Gorge attack. (US ARMY, 74265AC)

Kwangtung Province. Around the south, they had taken French-Indochina, the backdoor to China, and then moved up to Burma to seal off China from the west. There was now no way to supply China except from the air via India.

With the loss of the Myitkyina airfield (north of Mandalay) in northern Burma, the Allies lost their only intermediate stop from India to China along the southern route across the lower Himalayas. Japanese fighters based here would force any resupply efforts to fly a more northerly route from the Assam Valley in India to Kunming, China. The northern route required flying at higher altitudes and through severe turbulence and worse weather conditions. These were the longest supply lines of the war for American replenishment of equipment for the Allied cause.

They also passed through British territory and the resupply efforts for English troops operating in theater and further strained the tenuous connection. This was the situation that existed when Lt.

General Stilwell was appointed commander of the China-Burma-India Theater of Operations in February 1942. All three countries had been grouped together for the purpose of U.S. strategy. This was unfortunate because it seemed to imply an order of descending importance with regard to Allied policy. Burma had fallen, India was not at that moment facing a military crisis, but China was in desperate need of all the resources she could get.

> Following Stilwell's unsuccessful defense of Burma, where he had been in Command of the Chinese Fifth and Sixth Armies, he had hoped to train new Chinese ground troops in India. In addition to his military duties, President Roosevelt had instructed him to take charge of all Lend-Lease material in China prior to its delivery to the Chinese. This would enable him to wield considerable power in negotiations. But in the spring of 1942 the only CBI troops under his command were 3000 officers and men of the Tenth Air Force-mostly in India-and ninety-four ground personnel. (Samson, *The Flying Tiger*, 208)

The opportunity

There are times in the life of men when an opportunity presents itself and if taken at proper time, only good things follow. That was exactly the point that the Japanese field commanders found themselves in early in May 1942. Everything since their unprovoked attack against Pearl Harbor on December 7, 1941, had gone their way. Any undertaking that they had started seemed to be destined for quick victory. The war in Asia was going better than the general staff had anticipated. Hong Kong, Malaya and Singapore were now Japanese, as were the oilfields of the Dutch East Indies. Indo-China was a Japanese puppet state.

British and Americans surrendered at garrisons across the Pacific. The combined navies of the Australia, Britain, the Dutch and America had been decimated and forced to withdraw back to Australia for safety. The army forces facing the Japanese advance had been swept out of the way with little effort required. After some original surprise opposition by the British and American air units, Burma had fallen to the Japanese who were now poised on the borders of India and China. The Japanese now began to concentrate their forces for what they thought would be mopping up actions inside China itself, followed by the quick capitulation of the Free Chinese government.

Just as Hitler felt about the Russians outside of Moscow, Japan knew that they had only to kick the rotten door in and the corrupt Chinese government would collapse on itself. The stage was set for the undefeated Japanese Army and its air resources to begin a rapid assault into China to capture Kunming and successfully conclude the Chinese operations in Chunking. Victory would be theirs and then there would be hell for the white imperialist to pay. Banzai to the General staff, banzai to the Japanese Army, banzai to the Japanese Air Force.

At this moment, there did not seem to be anything besides the defeated, disorganized and the totally exhausted Chinese Army between them and victory. It was with great confidence, possibly with too great confidence that the Japanese Army launched their best troops to follow the 35 mile descent to the winding Salween River, through the magnificent gorges and down the winding road that led to the Salween Bridge at the bottom of the six thousand foot descent.

The first two days of this critical attack went like clockwork and every Chinese unit in the way was destroyed. Nothing the Chinese units fighting in this confined space could do seemed to slow down the Japanese 56th Division, the famous Red Dragon Brigade. At the rate that these elite troops were proceeding they would be across the Salween Bridge and at the top of the Salween gorge in two or three days. There at the top of the gorges the Japanese would rest briefly for a day or so while they reassembled their armies and then surrounded and swept past the hapless Chinese in Kunming and on to the wartime capital of Chunking.

In essence, the struggle would be over in a week, perhaps ten days at the most. The generals had already begun to draft the terms of surrender and to plan the pageantry of the formal Chinese surrender. Japan in early 1942 was on the verge of eliminating one of the Allied powers, changing the face of the war in at least three theaters.

This was the moment, this was the critical juncture. The 56th unexpectedly found itself ahead of schedule, well supplied with captured enemy supplies, at the border of China itself, and facing no effective Chinese resistance ahead of them. The famous Burma Road ran from Lashio all the way to Kunming, four hundred miles northeast of the city lay Chungking, China's wartime capital. All the Japanese had to

Figure 12–2. Tex Hill, Flying Tiger. David Lee Hill was trained as a Navy dive bomber pilot, then volunteered for the AVG. He was the squadron leader for the Salween Gorge operation, significantly contributing to the mission's success. (US Army, lonestarflight)

do was cross the Salween River, isolate and capture Kunming and then race unopposed across open, undefended ground to Chungking. There was absolutely no obstacle that could now impede their progress to final victory.

Well, there might be one small obstacle. One observer was in the ideal position to gauge the disposition of the Japanese Army and he also understood the significance of the strategic outcome should these forces reach Kunming and Chungking. Sitting high above the destroyed Salween Bridge, Flying Tiger Tex Hill watched as Japanese engineers rushed to complete a temporary structure that would allow the enemy onto the Chinese side of the gorge.

Shortly after the Burma Road entered Chinese territory, it crossed the Salween River via a bridge located at the bottom of a long, hazardous, winding road (see Map and Figure 3–1). This Huitong bridge across the deep gorge presented the only natural defense to entering the Chinese mainland. East of this point there were very few Chinese Army resources that were in position to slow down the major Japanese thrust into China itself. Before the Japanese assault into the Salween, the only effective Chinese forces of substance had been the 5th and 6th Armies to defend China's western provinces, since the bulk of fighting with the enemy forces deployed in the east and north. The 5th and 6th Armies were considered China's most effective force, but they had been seriously depleted fighting with the British in the Burma campaign. Their hastily retreating remnants were the only Chinese fighting force between the Salween and Kunming.

Recognizing this Chinese weakness, the Japanese quickly proceeded farther north in an attempt to attack and take Kunming. By May 5th the elite 56th Red Dragon Armored Division led the attack into the Salween River gorge as the Japanese Army approached the west bank. The road was a mass of confusion, damaged and destroyed Chinese equipment, masses of Chinese refugees and fleeing Army units completely clogged the one lane road and proved to be the only effective defense that the Chinese could offer. Too much stuff, scattered over too little space with too many terrified civilians had finally done what the armed opposition had failed to do.

The Japanese Army could not go forward until this mass of humanity moved from the road ahead of them. Like a cork stopper in the end of a bottle, the finest Japanese field army in China was stuck. It could not go forward because of massed civilians and they could not back up because of their 35-mile-long tail of troops and supplies to the rear. For the first time since the war began, the Japanese high command did not know what to do. While they sat and pondered their predicament, a solution was appearing on the horizon.

The other opportunity

Claire Chennault, the newly promoted US Army Air Corps Brigadier General and leader of the American Volunteer Group was acutely aware of the danger that the 56th Armored Division posed, since it had forced him to evacuate the bases in Burma from which his forces were flying their P-40 Tomahawks. He understood the strategic implications that these troops were bringing along with them as they went deeper into Chinese territory.

> Our pilots reported the Japanese motorized columns moving up the center of the road between swarms of unarmed Chinese soldiers and civilians on both sides. This was one of the few times during the war that I became greatly alarmed. If the Japanese got to Kunming, it meant the end of the war for China. With the Japanese in Yunnan the only possible supply route to China would be across the deserts of Turkestan and Mongolia from Russia. Cut off from Allied aid, China's resistance would collapse like a punctured lung. Japan would have a vast Asiatic base to exploit further operations into India and Siberia. This would come at a time when Russia was reeling under German thrusts to the Volga, the British were building a huge base at Karachi to receive the remnants of Montgomery's defeated Eighth Army should Rommel take the Nile Delta, the Japanese invasion of India seemed imminent, and the United States had all but lost the Philippines. (Chennault, *Way of Fighter*, 163)

This could have been the moment that the Second World War got away from the Allied cause. Everything was in place to prevent any possible salvation from this situation. Great Britain was on the verge of defeat in Africa and in the West, the United States did not have enough military forces to deploy to any of the other critical areas of the war. The mass economic potential of America had not had time to come into play-the American arsenal of defense was there, but it was empty. To every member in the Allied High Command above the rank of Captain the end seemed near. Fortunately for the Allied cause, there were an adequate number of low-ranking officers who were in just the right spot to change the history of the world that might have been.

> There was no effective resistance between the 56th Division and Kunming, and little from there to Chungking. China's fate depended on stopping the drive of the Red Dragons, for the great nation was already hard-pressed to the point of exhaustion. They would doubtless have forced the surrender upon the loss of either of these cities. China's capitulation, in turn, would have freed Japan to throw millions of soldiers westward into revolt-ripe India, just as the Germans were driving upon Cairo. It certainly would have changed the face of that campaign, probably allowing a link-up of the two Axis nations across India. What would have happened then, with the advent of the atomic bomb still three years away, can only be guessed. (Hill, *"Tex" Hill Flying Tiger*, 161)

A series of dramatic messages flew across the Allied radio frequencies as the Japanese forces approached the west bank of the Salween

River. The Chinese 36th Division and the 2nd Reserve Division-western China's last ground defenders-deployed on the eastern bank, having destroyed the bridge, when the Japanese reached the edge of the plateau on the west. Chennault sent an urgent telegram to Chiang Kai-shek on May 6:

> Latest reports say Japs on west bank Salween 1500 hours 5 May stop Bridge destroyed stop Japs meeting no opposition anywhere as soldiers civilians panic stricken fleeing east along road stop Consider situation desperate and Japs may drive Kunming in trucks unless roads and bridges destroyed and determined opposition developed stop Due to fact many Chinese trucks west of Salween presumably in hands of enemy request authority His Excellency the Generalissimo to attack between Salween and Lungling city. (Chennault, *Way of Fighter*, 163-164)

The true meaning of this communication reveals the desperation of the situation as Chennault saw it. He was asking for permission from the Chinese government to launch a last ditch air attack against everyone on the Burma Road, including tens of thousands of innocent civilians as well as Chinese soldiers. A strafing P-40 cannot discriminate between friend and foe, especially at speeds in excess of 300 miles per hour. Chennault knew this, that to save China he would have to kill thousands of his adopted countrymen, people he had come to love and respect, and that he knew loved him. These are the decisions, taken in the throes of battle that a commander has to make, that can't be taken back and that will never be erased from his mind. It was now that Chennault and Chiang made the decision and they knew that they would always bear the responsibility for what was about to happen. Old soldiers do what they know they must and now Chennault and Chiang did what they must.

> Generalissimo instructs you send all available AVG to attack trucks boats etc between Salween and Lungling city stop Tell AVG I appreciate their loyalty and redoubled efforts particularly at this critical juncture stop Please continue attacks especially boats and transports on Salween River stop Shall keep you informed immediately any change in situation. (Chennault, *Way of Fighter*, 164)

These messages went out in the morning and by noon on May 6, the Japanese had descended the twenty miles to the destroyed bridge. Japanese trucks, armored cars, tanks, cannons and many thousands of troops covered the entire route down and much of the western escarpment. With the Japanese stopped, they came under intermittent fire from several Chinese units that had maintained their ability and desire to fight. This skirmish continued for several hours as the Japanese waited for a unit of engineers and bridge-builders to arrive.

The mission

Chennault directed Tex Hill to choose three other Navy pilots to form a dive-bombing flight to attack the Japanese troops waiting on the west bank of the Salween River. Hill chose Ed Rector, who responded in his typical fashion, the plan was "simply marvelous," Frank Lawlor and Tom Jones, all ex-Navy dive bomber pilots. Tom Jones voiced concern that he hadn't done dive bombing since leaving the fleet.

> "Tom, it's a cliff wall a mile high," Tex replied with a grin. "You can't possibly miss." (Hill, *"Tex" Hill: Flying Tiger*, 158)

How could you refuse to fly with a leader like that? So off they went, eight P40's divided into two flights, four for dive bombing and four to fly high cover. The weather was terrible for takeoff with low clouds and intermittent heavy rain showers, in any other circumstances there would be no flying allowed. In route the flying was instruments only with heavy clouds and heavy rain, hard enough to fly in this kind of soup, much less to maintain formation, but fly it they did. When Tex estimated their time would put them near where the gorge should be, he began to let down through the clouds. Hill thought he spotted a break in the weather ahead. All eight P-40's broke through the weather and into the clear below the overcast.

> On a combat mission, it is rare to be captivated by the scenery; but this time, the pilots couldn't help it. Squinting in the sudden glare, they surveyed the purple-crowned mountains and the winding, storybook valley that cut its way through them. Sunlight painted the hillsides with rich colors, glittering blue off the tiny ribbon of the Salween below. (Hill, *"Tex" Hill: Flying Tiger*, 159)

Once again, the fickle gods of war came into play, just when and just where it was needed most, the weather broke and everything opened up for Hill and his little flight of P-40's. If the weather front

Figure 12–3. Huitong Bridge, Salween River Gorge. This bridge crossed the Salween River on the Burma Road to Kunming. During their retreat after suffering defeat in Burma in April 1942, the Chinese blew up this bridge to stop the pursuing Japanese. A large Japanese force waited on the west side (photo right) for their engineers to construct a pontoon bridge across the Salween River. Using heavy Soviet bombs on their new P-40Es, the Tigers set off landslides from the steep walls which blocked escape, then decimated the trapped Japanese troops in a series of dive bombing and strafing attacks. Photo shows an American convoy crossing after the area was recaptured and the bridge repaired in 1945. (US Army, 158-8)

had extended five miles further to the west he would not have been able to spot the Salween Gorge and its river below with its blown-up bridge.

As they descended to lower altitude it became evident that the entire road for almost thirty miles was covered with the Japanese Army and they had nowhere to go. They were trapped like rats in a barrel with no place to hide.

None of the pilots flying had ever seen so many Japanese and unbelievably there were no Japanese aircraft flying on scene. Hill dispatched Oley Olson, R.T. Smith, Erik Schilling, and Tom Haywood to fly high cover while he led his flight down to what he knew was going to be the most important turkey shoot of the war. It was as if divine intervention had placed Hill and his dive bomber wingmen above the Salween River for the express purpose of stopping the Japanese in place, just at the moment that should have given the invader the victory that they had been seeking for so long.

> Tex opened up the throttle all the way as he sent the P-40 into a sixty-degree dive. He knew the Japanese troops would be shooting all kinds of small arms fire at him as soon as he came into range; so, to make himself a more difficult target, he needed all the speed he could get. Meanwhile, he lined the P-40 up to target the top of the gorge, exactly where the road dived to begin the thirty-mile switchback down to the river. Gunfire began to flash below him as the troops opened up. It seemed as though the entire force stopped what they were doing to watch the Flying Tigers in amazement. At precisely the correct moment, Tex released his demolition bomb. The results were better than anyone had hoped. As he pulled into a hard right turn to watch, he saw the big Russian weapon slam into the cliff side, exactly where he had aimed it. Even at the distance he was, he heard the resulting whump. A cloud of dusty particles blossomed out into space. Through it, Tex glimpsed great slabs of rock peeling off the cliff surface and crashing onto the curves of the road below-and then, one after the other, his comrades lambasted the same area in turn. (Hill, *"Tex" Hill Flying Tiger*, 159)

This was the kind of mission every fighter pilot dreams about, but rarely does it come to pass. The Flying Tigers made pass after pass, up and down the gorge, all guns firing away. Fifty caliber bullets flying at near supersonic speeds, tearing through metal and bone, destroying everything that they contacted. This destruction went on unabated for nearly two days, until everything on this road of death was rendered useless. Japanese troops scrambled up the steep slopes of the Salween Gorge only to come tumbling down after another pass of the P-40's.

Finally, a merciful god decreed that enough damage had been done to the Red Dragons and the rest of the once mighty Japanese Army that had descended down to the banks of the Salween River. Those lucky few that had survived the ordeal of that bloody forty-eight hour carnage managed to crawl, or walk or run away, back to where they had come from. Back toward Burma and India, never would they come this way again. What should be known as the

Miracle of the Salween River was now done. It should be ranked along- side of the Miracle of the Marne from the First World War for its significance. Just as at the Marne, the Miracle of the Salween had changed the course of a gigantic struggle that would once again save the world from the evils of dictatorship, oppression, and fanatical empire. The world had come to the brink of destruction and could now step back because of the bravery, skill, and determination of a mere handful of American volunteers. The world was not safe for democracy just yet, but it had come a long way in the last two days.

Back from the brink of disaster came Free China, back from the brink of disaster were those few Chinese units that stood their ground and fought at the banks of the Salween River. These unknown, forlorn troops that fought and died against overwhelming odds until eight tiny specks appeared overhead and performed yeoman's service for China. The Japanese invasion did not just stop, it was destroyed on that road down to the Salween River Bridge. No one yet understood the true significance of what had happened in those forty-eight hours. One of the most important battles of the Second World War had been fought and won by China and a handful of American pilots. Let the Japanese celebrate all their victories to date, the one battle that they had to win, they lost. Chennault had an inclination to believe that the most significant battle of the Asian War had just been fought and won by the Allies but it was still too early to know.

13

The End of the AVG

American military involvement in China

As the American military geared up for war in Europe and in Asia, the European threat seemed greatest.

> The loss of Burma, completing the blockade of China, raised the fear in American minds that China's will to resist would not survive her isolation... Priorities in any case were assigned to Europe. Strategy in Asia was still uncertain. All that could be envisaged so far was the necessity of keeping China in the war as a base for air operations against Japan's sea lanes and ultimate springboard for invasion. "Keeping China in the War," the title of a War Department plan at this time, meant in effect sending her enough supplies to keep her operational. (Tuchman, *Stilwell the American Experience*, 301)

Army officers began arriving in China to direct the American war effort in Asia. Their headquarters was set up in Chunking, and they all needed American food and office supplies flown over the Hump.

Change is coming

Chennault knew the conventional American military establishment coveted the AVG. Army leaders did not want him as the commander of a regular air corps unit. His past reputation, non-conformity, insubordination and aggressive personality returned to haunt him. If or when the volunteer group deactivated, Army brass wanted Chennault replaced. Old grudges came to the forefront. These men were less inclined to follow Chiang's leadership. Any chance for effective operations in China vanished without Chennault's presence.

It is ironic that Chennault, the man considered a loose cannon by senior American officers, was the only successful Allied commander in Asia at the start of hostilities. The established military sought to remove their most experienced commander from the only successful American fighting unit in China. This would certainly destroy the AVG's cohesion and effectiveness. Chennault's success was due to independence of command. Working under Chiang, Chennault was free to respond to any unexpected situation that arose. Flexibility to circumstances as they happened was a key factor.

Regular Air Corps units are bound by procedures and regulations. By nature, these restrictions prevent quick response and diminished effective counter- measures. The brass were killing the goose that laid their golden egg. If the ring knocker faction succeeded, Chennault would be removed from command. If that happened, China would be finished as an effective ally. No other officer taking over the AVG could match his success in military or political spheres.

> It became Chennault's litany that a small but effective air force, free to strike at the most advantageous time and place, could produce results far out of proportion to its strength. He saw the opportunity. He raged inwardly at not being able to exploit it. (Byrd, *Chennault*, 139)

To no one's surprise, the official reply to his suggestion was no. Ring knockers determined that such a small force was incapable of producing his claims. More braggadocio from an overconfident and outspoken Chennault. Those who never fought the Japanese again doubted the only man who brought victory to the Allied cause in Asia. Even after all these years, it seems incredible that certain senior military officers in time of war would not consider ideas from a man they disliked. Negligence and slander proved a heavy cross for Chennault to bear. Prejudiced ring knockers would rather sacrifice untold American lives in combat than admit Chennault was right.

The road to perdition

United States government officials commenced secret negotiations with China in mid April 1942.

> In the course of negotiations with Chiang Kai-shek over the reabsorption of the AVG into the American Army, Marshall had promised him that Chennault would be the ranking air commander in China. (Tuchman, *Stilwell the American Experience*, 248)

However, on the last day before General Marshall left China, the situation was still unresolved. Air Corps personnel wanted command of the AVG and induction as a complete unit. The Air Corps leadership wanted to get rid of Chennault all together. If that proved impossible, then they wanted to make sure he was not in a position of command.

> Chennault was now raising loud objections on the grounds of the broken promise although his real objection was serving under the officer designated to be his superior, Colonel Clayton L. Bissell, with whom he had an old feud. Everyone entered the quarrel: Stimson was "quite concerned," Stilwell was sent for, Currie intervened on Chennault's behalf and the Air Force chief of staff, General Hap Arnold, who shared the conventional officers' dislike of Chennault as a black sheep, "hit the ceiling."...As overall commander of Americans in the theater, Stilwell insisted that Bissell rank Chennault and Arnold so ordered. (Tuchman, *Stilwell the American Experience*, 248)

Stilwell wanted Chennault under his thumb and acted:

> It is recommended that Col Clayton L Bissell O-10474 be promoted to brigader general Army of the US and that Col Claire L Chennault be promoted to brigader general Army of the US, effective date of Chennault's rank to be one day subsequent to the appointment of Col Bissell. These recommendations have the concurrence of the generalissimo. (Stilwell, *Stilwell Papers*, 78)

Herein lies the reason for all the rancor and distrust in China until the end of the war. Chiang and Chennault never trusted the Army High Command again. Chennault felt betrayed by the ring knocker conspiracy and Chiang seldom believed the promises of Roosevelt or any American leaders because of this incident. Lack of trust led directly to the disaster that befell China during the closing war years and afterwards.

Chiang Kai-shek negotiated a deal he could live with and part of that deal retained Chennault as commander under his authority.

In addition, the Chinese wanted AVG personnel to form the nucleus of whatever fighting force was deployed to China. Promises made to Chiang during these negotiations were meaningless. Seeds of further distrust were sown here.

> In Washington, however, negotiations hit a snag, possibly precipitated by reports the AVG displayed poor discipline or none and that the men were not eager to accept a reduction in pay. (Byrd, *Chennault Giving Wings to the Tiger*, 141)

Senior officials who saw members of the 1st AVG in India or Africa found them disorderly, disrespectful and completely unmilitary. These charges were mostly true. After their year in Asia under Chennault, it wasn't possible for them to adapt to military regulations. They were free of discipline for too long.

The real reason for the snag was that Stilwell refused to recognize the merits of Chennault and the Flying Tigers as patriotic fighting men. He considered them paid mercenaries, only a step above pirates. Stilwell believed (correctly) their value to any structured military organization was minimal. This was an unacceptable position for the Generalissimo.

To guarantee this strategy required huge investments of American personnel and material. These assets in place, Stilwell assured success in less than 10 years. Chennault and Chiang voiced strong opposition to a flawed, unnecessary campaign. Stilwell thought Chiang and Chennault were playing politics. They were only interested in feathering their own nests. Stilwell refused Chennault as the ranking U.S. air officer in China, contrary to promises made to Chiang by President Roosvelt. No compromise on these issues was forthcoming, so Stilwell's plan came to naught.

> Just when it seemed like nothing could get worse than it was, the tide began to turn. Chennault was urgently summoned to Chungking by the Generalissimo to attend a military conference between the U.S. Military Mission and the Chinese Government. When he returned his face was pale, but his black eyes flashed with excitement. Then the cables began arriving. Captain C. L. Chennault returned to active duty and was promoted from Major. Major C. L. Chennault was promoted from Major to Colonel, effective 9 April 1942. Colonel C.L. Chennault promoted to the rank of brigadier general, effective 21 April 1942. Perhaps the most rapid advancement from Captain to Brigadier General in the history of the Army Air Force. (Trumble, *Biography)*

Figure 13–1. Chennault with the Generalissimo and Madame. Photo likely taken at Kunming in April 1942, during a stop to visit with Chennault. They continued on to Burma to see General Stilwell. (Neale, 126)

In retrospect, this was one of the most tragic events in the long history of China. Everything averse to American policy that has happened in China is traceable to this unfortunate choice of command. All trust between Chiang and the United States stopped.

Within the military of 1942, there were only two men knowledgeable and experienced enough to deal with the difficult conditions. Whoever was selected must work with the Chinese government and the existing state of military affairs. On the surface, both were qualified to hold the senior military command.

On one hand, Chennault was considered a renegade, an outsider and troublemaker by the U.S. military establishment. Not a graduate of West Point and never a team player, his qualifications for command were disputed. Chennault's main strength arose from his defense against the Japanese, successful dealings with the Chinese government, positive relationship with the leaders of the Chinese military and his loyalty to Chiang and China.

Chennault did a masterful job working with the Generalissimo and Madame. Facing many difficult military and political crises, Chennault always found satisfactory solutions for problems that arose. He maintained the highest esteem of the Generalissimo and Madame and held their complete trust. Both knew Chennault's love of the Chinese people was genuine. He proved both his genius and loyalty many times. Chennault was the only American ever given this honor, and he never disappointed Chiang or Madame.

Stilwell selected newly promoted Brigadier General Clayton Bissell overall theater air commander. Bissel's promotion date was purposefully made one day before that of Chennault. Newly appointed Brigadier General Claire Chennault must therefore work as Bissel's subordinate. Bissell was a World War One ace, Chennault's instructor in flight school and his nemesis at ACTS. Additionally, he was overtly arrogant and a proponent of bomber superiority. Bissell was an original member of the bomber mafia. He was a longtime opponent of Chennault. The two men did not like each other. Nothing good came from this working arrangement.

> As the successes of the AVG mounted, the U.S. Army Air Corps began to realize that, sooner or later, they would need an air force in the China-Burma-India Theater. As they considered who its leader should be, Chennault's name surfaced; and immediately, jealousy among the Air Corps brass reared its ugly head. (Hill, *"Tex" Hill Flying Tiger*, 142)

Once again, we see that Chennault is unable to free himself of his past reputation. The ring knocker conspirators would not even give him credit for his spectacular success in China.

> There was a period of almost three months that I didn't see Chennault. I learned that he was down with bronchitis and having a miserable time of it, and also I knew that General Bissell's interference with Chennault's plans for aerial combat in eastern China were playing havoc with the OM's temper and control. Long before the AVG disbandment and Bissell's taking command of the Tenth Air Force, the General had told me with contemptuous amusement of Bissell's one-time idea of bombers dropping weighted metal balls into fighter aircraft propellers if they came close enough. There were no secrets in the American air forces in China about the relationship between the two. Next to the Japs, Bissell was the most hated man in China. (Trumble, *Biography)*

The long-standing animosity between Bissell and Chennault went back to the 1920s and grew more acrimonious over the years. This was a costly, uncomfortable command structure. It had detrimental effects for China and Chennault. If General Marshall was trying to sabotage the relationship between China and the United States, he made the right choice.

Figure 13–2. Brigadier General Clayton Bissell. Bissell devotedly attempted to undermine Chennault's authority with the CATF and the 14th Air Force. The harm to China's war effort was substantial. (www.af.mil, in Biographies)

> Upon the rout of his forces in Burma, Stilwell set out on foot to traverse the hundreds of miles of jungle, rivers, and mountains that lay between him and India. The retreat lasted weeks; and through it all, the general's personal courage and toughness were incredible for a sixty-year-old man, as he led the one hundred fourteen souls in his band to eventual safety...Stilwell refused a transport plane that arrived to lift him out, flown by Robert Scott and Caleb Haynes, and further declined to give a single report to the Generalissimo during his exodus. "I doubt he knows the meaning of the word discipline", Chiang wrote angrily in his personal correspondence. (Hill, *"Tex" Hill Flying Tiger*, 161)

Dysfunctional senior command structure guaranteed acrimony and dissent for the duration. Chiang and Stilwell completely detested each other. Neither cooperated nor respected the other. Ensuring even worse behavior, both subordinate commanders Chennault and Bissell were not compatible and never worked efficiently either. Both sets of commanders were the antithesis of the other. This did not portend well for harmonious in-theater operations. The system imposed on Chiang and Chennault could never produce victory. Most certainly, it had the potential to extend the war in China.

Command is never an easy thing, sometimes it is an impossible thing. If the Japanese were lucky, they might just steal victory out of the jaws of potential defeat. These were the facts of life and Chennault had to deal with them. It would remain this way until either he or the war came to an end.

The real irony is the way Chennault the maverick, the officer George Marshall and Hap Arnold considered a loose cannon, overcame his deficiencies. Over the last five years Chennault reinvented himself or rather he was allowed to be himself. Like the Phoenix he rose from the ashes of his former career and was the darling of the Chinese. Chennault was accepted into the highest levels of Chinese society. He was considered a sage among his peers. Genius, reliability and dedication endeared him to the Chinese people. Chennault thrived over the adulation.

> [In 1937,] When the Generalissimo realized he had to have an air force...he appointed his wife, Madame Chiang Kai-shek, to be Chairwoman of the Commission on Aeronautical Affairs. By then General Chennault had made his abilities known and his loyalties known and the Gissimo appointed him Chief Advisor to the CAF under Madame Chiang. (Trumble, *Biography)*

Matters of government, military planning and operations were his strong points. His word was gold. Chennault proved time and again his loyalty to Chiang and the people of China. Chiang and Madame enjoyed his complete loyalty.

> Bit by bit, piece by piece, the picture of General Chennault's relationship with the Generalissimo and Madame Chiang Kai-shek and his role in creating an effective Chinese Air Force, became clearer in my mind. No other Westerner except W. H. Donald was held in more esteem than Chennault at the time he was ordered by the Generalissimo to create and command an American Volunteer Group. I never met Mr. Donald. At the time of the Japanese invasion he had been taking a rest in the Philippines, and had been interned in Manila. (Trumble, *Biography)*

Chennault believed at heart he was Chinese, not a foreign devil. Of the thousands of Americans fighting in China Chennault was the one foreigner trusted by the Chinese.

> With the uncanny ability of the Chinese to know whom to trust and whom not, Madame put her entire trust in Chennault to create and train an effective Chinese Air Force. Together they had mapped out a long range plan to train pilots, construct strategic air fields, obtain American planes and set up the logistic organization to operate effectively. While not exactly being only a figurehead, General Chou's role at first lay in creating an echelon of command and organization in line with Chennault's ideas. And to man the various airfields with logistical and maintenance personnel and to see that they were properly supplied; also that they were properly embraced within the telephone and radio communication air raid warning network which had so baffled the Japanese efforts to make surprise attacks. (Trumble, *Biography)*

Now a committed team player, as commander he was in complete control. Chennault worked within channels when permitted. Senior air corps officers (ring knockers) believed this was impossible for Chennault the Maverick. Maybe they just did not want to know. Wrong words in the right ears now proved irrevocable. Unfavorable impressions by the American "old-boy network'" continued working against Chennault and China.

> Despite Chennault's apparent success in fighting the Japanese, he was the subject of intense discussion in Washington after a year of war in the dying days of 1942. On December 16,1942, General Henry H. Hap Arnold, chief of the Army Air Force, was briefed by an unidentified member of his staff on the situation in regard to the 10th Air Force and the strained relations between

> Chennault, and Generals Stilwell, commander of all forces in the C.B.I., and Clayton Bissell, then commanding the 10th Air Force. It was a battle between vastly dissimilar personalities and military philosophies. (Glines, *Forgotten Warriors*, 20)

The fact that Chennault and Bissell were devout enemies was well-known throughout the Air Corps. That the choice of commanders in the CBI did not present any chance for success.

> Here are notes from that secret briefing: Generals Stilwell and Bissell reported that Chennault had five categories of personal shortcomings: [1] lack of administrative ability; [2] lack of appreciation of supply problem; [3] lack of discipline; [4] poor judgment in selecting key personnel; and [5] a tendency to take the part of the Chinese in controversies between General Stilwell and the Chinese. (Glines, *Forgotten Warriors*, 20–21)

This report seems to bear the fingerprints of Clayton Bissell. Without faith or trust in Chennault's ability, there was no chance for smooth operations in the CBI. Chennault's old enemies tried to change him from hero to goat. Our anonymous reporter continues,

> General Chennault has difficulty adapting himself and his former AVG personnel to Army methods. He and they reason that by employing their own methods, they in the AVG and in the China Task Force were consistently able to defeat the Japanese in every encounter...Theirs is the easy, happy-go-lucky way. Army discipline and methods employed by Generals Bissell and Stilwell are distasteful, and are considered unnecessary red tape. (Glines, *Forgotten Warriors*, 21)

The report cuts to the crux of Chennault's problem in China. It is a perfect encapsulation of the ridiculous. This questionable work represents a contemptuous picture of Chennault. It is demeaning at every point. No attempt is made to check the accuracy. Allegations, all untrue, arrived steadily from members of the ring knocker conspiracy. Callous and uncaring the masked writers casually cast off all his accomplishments. With the wave of a hand five years of exceptional success were negated.

> Intelligence is a gift from God; patience, determination, gentleness and understanding, self-confidence and mastery of a human art, are the products of self-effort. Possessing these things, Chennault captured the affection and loyalty of nearly everyone who served under his personal command. When his competence and knowledge of aircraft and pilots and aerial combat were first

> demonstrated to the satisfaction of the Generalissimo, it became his fiat that Chennault's instructions or "suggestions" be followed to the letter. If not, the penalty would be immediate execution. That was in 1937. In 1941, when I became his secretary and through the next five years, it was my experience that there was no longer a need for the G-mo to enforce the General. Most members of the Chinese Air Force had full confidence in him and cooperated with him fully and willingly. (Trumble, *Biography)*

The American High Command remained satisfied with their negative assessment of Chennault. Those who did nothing criticizing those who did everything. This is how the old boy/we know better than you network works. Woe be to you if this network is ever against you. Saddest of all was the readiness of most senior officers in the Air Corps to accept these unwarranted, untrue allegations. Because of past sentiments they held preconceived ideas not based in fact but pettiness. Our anonymous soul feared he might just run into Chennault in a dark alley sometime. Better to remain anonymous than sorry. Overlook the facts as long as that supports your own ideas.

> Adding fuel to that dislike was the news of the constant setbacks that our forces in the Pacific were experiencing. The Japs were having it all their way and outside of Chennault's air strikes against Hankow and river traffic, there seemed to be no war against the Japs as far as China was concerned. It seemed like both sides had dug in and neither wanted to break through the others lines. (Trumble, *Biography)*

> With this parting remark, our secret reporter added, "It is generally known that the British officers in India and Burma did not highly regard Chennault and the AVG. In the eyes of these conventional people, the AVG was a motley organization. One British general described the AVG as louts [who] even paint nude women on their aircraft." (Glines, *Forgotten Warriors*, 21)

Oh my word! Nude women painted on airplanes! Have you ever seen the nude women they painted on their aircraft? These naked women look like stick figures and are not controversial, especially when compared to nose art in England. There is no end to their pettiness.

This reeks of T. E. Lawrence, another unconventional commander fighting a guerrilla war. What loutish brutes. They go together so well. This from the acknowledged masters of warfare, the only military professionals in Asia, those defeated at Hong Kong, Singapore, and Burma. If Chennault was going to listen to anyone, it would be Japanese. At least they knew what they were talking about.

By the spring of 1942, the men of the 1st AVG heard rumors they were being dissolved and inducted in some form into a U.S. military organization when they returned home. Learning of Chennault's promotion to Brigadier General, they were happy for him. The men, however, considered the government position toward them very prejudicial and unfair. They realized their fate was sealed and presented them with very few options.

Many of the Tigers wanted retention in China under Chennault's command. Most enjoyed their service in China. They considered themselves still members of the regular Army Air Corps. Those ordered unofficially to join the AVG felt they only did their duty when they joined up. These men were proud of their service to date and had fine military records. They were patriots. The Flying Tigers did the job they were ordered to do, even if it was in secret and done with a wink and a nod.

How to win friends and influence people

The 1st AVG fought a difficult war in Asia. They deployed away from family, country and the comforts of American life for almost a year. One pilot said of his time in the AVG,

> This was the beginning of the greatest adventure I ever hoped to experience. It wasn't until years later that I fully realized the magnitude and significance of this first step to be a lifelong adventure into the mystic Far East. (Life Magazine article, *April 1942)*

All wanted to go home on leave before returning to the war. They were recognized heroes and icons of American aviation. Fresh from the battlefield, they deserved thirty days of home leave. After home time, they would gladly return to the war.

These men were patriots recently rendering invaluable service to the war effort. Among the best fighter pilots in the United States military, they wanted to stay in China and fight under Chennault. The Tigers admired, even loved, their commander and wanted to see out the fight with him.

Pilot R.T. Smith was asked, "Did you ever regret joining the AVG?" He glanced off to the side, put his tongue in his cheek and said: "Only on those times I was being shot at." (Life Magazine article, *April 1942)*

With this type of modesty, it is easy to see why most of the men wanted to stay under his command in China.

These men were serving their country. The Tigers wanted to stay with their leader for the duration. What could possibly be wrong with this idea? Their requests were common enough and were simple and easy to fulfill. All they asked for was a regular commission in their branch of service. They wanted priority travel to go on home leave. The men were a unit proud of their service and wanted to fight together again. These were reasonable demands, and they were readily given to others. It was no big deal. I suspect it was petty jealousy from 10th Air Force headquarters acting against Chennault and, by association, any of his men. Somehow, this doesn't seem to be the right way to treat heroes.

In view of everything they accomplished and their excellent flying records while fighting in China, they deserved the right to go home on leave. This was a reasonable request. The distance of seventy years did nothing to abate their righteous indignation. Compliance was only a matter of paperwork, nothing difficult about that. Common sense recognized this was appropriate. What else could the Army do but say yes?

The answer when it came was something else. "Request Denied"! Well, as Chennault was fond of saying, "that really set the cat amongst the pigeons." No good would ever happen between the Flying Tigers and the United States military again. Nothing but feathers and a—holes remained by the end of that day. The former from the Tigers and the latter from the brass of the Army Air Corps.

The Flying Tigers were not an official military organization contained within the command structure of the United States Army. No way, no how. In the circumstances of their formation they could not be. Regular air corps troops that saw them in India or Africa thought they gave a ragtag appearance of mercenary soldiers. Messy uniforms of mixed parts, unshaven and unruly. Professional Army types could not believe their eyes. The Tigers looked more like pirates than military men. They were slovenly, unkempt, with no sense of military bearing at all. Uniform dress codes went out the window. Pistols at their side and swagger in their step, these men would never pass as military.

Army high command (ring knockers) thought these men would never be acceptable. Not part of their tailored uniform, spit and polish officer corps. Professional military personnel were aghast when the Tigers walked irreverently through airport lobbies or base operations. Who or rather what commanded these misfits? Members of the 1st AVG could be in the French Foreign Legion if one went by uniforms alone.

Their seemingly anti-military style was not conducive to regular Air Corps discipline, especially at soft headquarters billets in India and Africa. No, those guys would never be acceptable! When the rear echelon brass discovered these outlaws belonged to Claire Chennault no one was surprised.

Word trickled down through that invisible, all knowing and sometimes correct grapevine. The Tigers were disbanding. Chennault didn't know it yet, but his troops were on the verge of mutiny with the Army brass or at least pistols at twenty paces.

Bissell's mistake

Fearing something like this was in the air, the official military tried to smooth things over by sending their new public relations specialist, Brigadier General Clayton Bissell. Stilwell's senior Air Officer, the man selected as their boss, arrived from India to speak with his new charges. Remember, this man Bissell was new to the Asian Theater. He was not up to speed about who he was addressing. The Flying Tigers were an enigma to him.

> Bissell confidently requested that Chennault allow him to speak to the men himself. The Old Man didn't think it would help the situation, and for his part, he thought the boys deserved a furlough to the States; but Bissell was insistent. Thus it was...that most of the men gathered in the auditorium at Yunnan University to hear what the general had to say. From the outset, Bissell's speech was an unmitigated disaster. He began by downplaying the AVG's accomplishments as negligible–adding that the Western press had fabricated and enhanced their stories to "make them look good." (Hill, *"Tex" Hill: Flying Tiger*, 168)

Hearing the new boss's speech Chennault realized how things were going to be. He never cared for Bissell as a person. Now listening to his harangue to men he hoped to lead in battle Chennault remembered why. Bissell dismissed their brilliant performance as hype by the news media and overblown. Nothing would ever follow but mutiny and murder. Chennault realized this man could not lead a herd of thirsty cows to a watering trough. Bissell did more damage in one hour to the morale of the finest Allied air unit in Asia than any air army of Japan. Now started a very long, exasperating time for Chennault. Well, he thought "maybe one of the boys might just shoot the bastard."

Bissell's "Word from on High" came to the men of the 1st AVG. Once disbanded, they would be ordered to return to active duty immediately. No leave or priority transportation authorized. Their only option, rejoin their original branch of service or stay in the Army Air Corps, but as reserve officers without permanent commissions. They faced swearing in and induction into the U.S. military in China on the fourth of July 1942. If they refused, Bissell guaranteed immediate induction upon arrival in the States.

Induction of trained pilots and seasoned ground personnel into the Army as privates was not a viable option. The Flying Tigers considered this very prejudicial treatment and unfair considering their outstanding war record. Many were Reserve Officers nearing the end of their four-year enlistments, and getting a permanent commission was important. Bissell behaving like an asshole was like bad icing on a bitter cake.

As it became apparent that few in the AVG would join, Bissell privately relaxed his "no permanent commissions" policy for selected individuals. Tex Hill, now Squadron Leader of the Panda Bears, missed the big meeting because he was ferrying a new P-40E across India. He wanted to continue in China, and Bissell made him a Major and Squadron Leader in the new organization.

> Only four pilots besides Tex had agreed to accept induction into the Air Corps and stay in China indefinitely: Gil Bright, Charlie Sawyer, Frank Schiel, and his old friend Ed Rector. Twenty-two ground personnel had also opted for induction. By doing so, they would provide a small but vital core of experienced leadership... (Hill, *"Tex" Hill Flying Tiger*, 173–174)

Bissell never understood the resentment from the rest of these men. Why these new flyers disliked him was a mystery. Why wouldn't they want to serve under his leadership?

China Air Task Force

With the dissolution of the Flying Tigers, what would be the organized, trained air unit in place to fight the coming air rush from the Japanese when the monsoon ended? The 10th Air Force headquarters personnel believed (so they claimed) from their paperwork that adequate resources were on hand. The new China Air Task Force would consist of the 11th Bomb Group (flying B-25 bombers) and the 23rd Fighter Group (flying P-40s). The former 1st, 2nd, and 3rd squadrons of the AVG would become the 74th, 75th, and 76th Fighter Squadrons. The fourth of the four squadrons authorized for the 23rd Fighter Group was still to-be-announced.

> Stilwell solemnly promised the Generalissimo to replace the A.V.G. with a full-strength Air Forces fighter group of four squadrons and a hundred planes. On the delivery date set for them-July 4th, 1942-the Generalissimo found himself exchanging the veteran A.V.G. for three newly activated squadrons of the 23rd Fighter Group that existed largely on paper. Actually the Army supplied only a dozen green pilots and twenty clerks and mechanics. (Chennault, *Way of a Fighter*, 176)

Chennault realized the true gravity of the situation. Something must be done to rectify this and it must be done soon! As a young farmer back in Louisiana, he learned never to waste anything, how to recycle, how to improvise. These traits served him well while commanding the 1st AVG. Remnants left behind were the only tools Chennault would have to fight the Japanese while men and equipment trickled in. The remaining aircraft from the AVG had to continue flying: 31 well-used 51 P-40B models and 20 newer more capable P-40E arrivals.

> The China Air Task Force was patched together in the midst of combat from whatever happened to be available in China during the gloomy summer of 1942. There was precious little. As the stepchild of the Tenth Army Air Force in distant Delhi the C.A.T.F. had to fight, scream and scrape for every man, plane, spark plug, and gallon of gas. (Chennault, *Way of a Fighter*, 176)

Operating as the most distant part of the longest supply line of the war guaranteed Chennault and whatever they called his men the least and last of everything. Stilwell and Bissell made fighting very difficult, but it was the shortage of supplies that was Chennault's real enemy. Army ground equipment, promised immediately to the 23rd Group, reached India one year later in the summer of 1943. Vital equipment wasn't unpacked and only showed up in China the next fall.

Taking stock of his war chest, Chennault realized he was losing most of his pilots on July 4th. He appealed personally to the AVG members to stay an additional two weeks to train newly assigned crews. Eighteen pilots and twenty-one ground crew volunteered, mostly from Neale's 1st Squadron and Hill's 2nd Squadron.

Bob Neale, senior AVG squadron leader, volunteered for the two weeks and assumed temporary command of the 23rd Fighter Group. His replacement was Colonel Robert L. Scott, a most capable officer. Scott had flown several "guest" missions with the AVG earlier in the year. The three squadrons of the 23rd were commanded by former AVG pilots Tex Hill, Frank Schiel and Ed Rector, newly promoted to the rank of Major in recognition for their sterling combat records.

Attached versus assigned

Since formal Army assignment of pilots or equipment to the CATF was so "delayed", Chennault resorted to a process called "attachment" that had started with the AVG. Personnel in other Army units could be temporarily attached to the CATF for a specific purpose, by mutual agreement with the other unit.

For example, Colonel Scott, although assigned to the Air Transport Command flying supplies over the Hump from India, was flying impromptu P-40 missions from India over Burma. Scott met Chennault in Kunming during his transport missions, and pestered him until he relented and let him fly with the AVG. When Chennault needed someone to command the 23rd Fighter Group, Scott was attached to the AVG and flew 9 combat missions with the AVG. This way, Chennault could see how well Scott fit with the men of the AVG. This led to Scott's selection as head of the 23rd.

Figure 13–3. Tex Hill with Attached Pilots. Left to right: John Alison, Tex Hill, Albert Baumler, Mack Mitchel. The men are standing next to a P-40E with a new logo for the Flying Tigers: an Uncle Sam Tiger in China. (US Army, 73991AC)

Another example was how Alex Baumler ended up in the 23rd. He had signed a contract to be the 100th pilot in the original AVG. However, the State Department refused him a passport because he had flown combat missions in the Spanish Civil War in defiance of State Department restriction. He returned to the Army, which eventually made him a liaison with the AVG. As a trained combat pilot, he was attached to the 23rd, flew missions with them and eventually became squadron commander of the 74th.

This process reached its zenith when Chennault more or less stole the 16th Fighter Squadron of the 10th Air Force, 51st Fighter Group from India's Assam Valley. John Alison led the first flight from Assam to Kunming June 27th, where they were stationed at Peishiyi near Chungking.

> By inviting a single flight at a time to China for "experience" I lured the entire squadron. They were bogged down in the monsoon anyway, to Peishiyi during June and July, and [I] never returned them. The 16th was well equipped with Kitty Hawks [P-40E] and veteran pilots. ... The 16th stayed on in China until the end of the war and fought through all the hottest actions along with the 74th, 75th and 76th Squadrons of the 23rd Group. (Chennault, *Way of a Fighter*, 176)

Chennault was proud of his feat of deceit. He told the story many times over the following years.

> Granddaddy was fond of saying we were fighting the same enemy. Those guys did us a lot more good than where they were in India. (Nell Chennault Callaway, *conversation with author)*

Casualties

Voluntary extension by members of the AVG led to further casualties. Tom Jones of the Salween action crashed to his death while practicing dive-bombing. Bob Little died attacking Japanese positions on the Salween River. Newlywed John Petach died in combat during the final days. Fourteen extra days came at a high cost to the Flying Tigers. Their sacrifice protected the newbies from a vengeful enemy and gave the 23rd Group some much needed experience and leadership.

When the two-week extension was up the last of the volunteers began their homeward journey. Charley Bond took in-tow newly widowed AVG nurse Emma Jane "Red" Petach for the long trip back to the States. She never faulted anyone for her husband's extension that cost him his life. Secretly pregnant and still grieving her loss, she knew her husband John died doing what he wanted. He was a warrior serving his commander and China when needed most. For this and many other reasons, Tex and the Flying Tigers will never forget Red Petach or her unborn daughter, as long as any of them are still alive.

With the demise of the Flying Tigers at hand, one of the most fortuitous actions of the war in Asia came to a close. Without the combat leadership and courage of Chennault and the men of the 1st AVG it is difficult to see how China, in spite of her best efforts, could have successfully fought the Japanese. If China didn't remain free (that was a real possibility) the war in Asia might be lost before America became fully engaged. The people of China recognize what Chennault and his brave men accomplished even to this day.

Chennault and his AVG staff kept in contact with families of dead or missing personnel. This was a heartfelt action on their part. Chennault owed every man in his command this service. Both Olga Greenlaw and Chennault corresponded for years with all of his boys and their families, providing any news they gathered.

> My Dear Mrs. Meadows: I was very pleased to receive your letter and to hear the good news that Betty will come home. I have been awfully worried about her and it is a great satisfaction to me to know that she is alive. Scotty and I used to talk about her for hours and we were always planning some wild scheme by which she could be rescued. We hoped she would be sent to Shanghai and from there Scotty had a good idea of how he was going to take her away to Kunming, where I was stationed. Scotty would rest in peace knowing that Betty was safe and back with her parents. God grant that this will be so. Scotty is buried in the military cemetery in Kunming, not far from the airdrome. He had a Christian funeral. I remained in the cemetery until his casket was lowered. The cause of the accident was motor failure. No one was at fault except the plane was too worn out and overloaded, as usual. I am very happy that you wrote to me and would like to see you someday. Scotty was very dear to me. I always felt that he and Betty were a brother and sister of mine. Poor darlings! (Olga Greenlaw, *letter to Mrs. Edward O. Meadows*, 1943)

Madame throws a party

Now it was over, the most important action many of these men ever undertook. Their service was vital, their courage magnificent and their devotion to America unquestionable. Now greatness slipped quietly into the past.

> On the evening of July 3, the Flying Tigers gathered at the estate of Lin Sen, the elderly president of China, in Chungking. H.H. Kung, Madame Chiang's brother-in-law and China's Minister of Finance, was "laying on" a party for the Tigers. The food was excellent, the facilities breathtaking; but somehow, the melancholy drizzle outside reflected everyone's mood.... They were about to lose something, and they could feel it slipping away. When the clock struck the hour of midnight, the American Volunteer Group passed into history. (Hill, *"Tex" Hill Flying Tiger*, 176)

History recorded their deeds, and all did their best. The Flying Tigers fought in a glorious cause and suffered the loss of comrades and friends. Most never outdid what was already done. It was with real sadness they went home that night destined to be but a footnote in history.

> I will think of all of you during your 4th of July reunion and pray that God blesses and keeps my "Angels" under his wing. Please give my fond regards to them, and tell them that the Republic of China and I will always remember and appreciate their valiant contribution. (Madame Chiang Kai-shek, *Letter*, 2 May 1985)

It is said that whenever a successful general returned to Rome after a great victory, a parade was held in his honor. The citizens turned out in tribute to his success. The victor rode ahead in an adorned chariot to the adulation of all of Rome. Behind him stood a slave whispering in the general's ear, "Fame is fleeting." This was now the lot of the Flying Tigers—fame was fleeting!

> On the 4th of July, 1942, the sun rose on a new organization. The China Air Task Force was born, cobbled together from newly arrived Army Air Corps personnel and the long-suffering remnant of the Flying Tigers who had agreed to stay...On the day it came into existence, the strength of the CATF was scattered in various places. The 74th Fighter Squadron, led by Frank Schiel, was at Kunming; Tex Hill had been given the 75th Fighter Squadron at Hengyang; and Ed Rector took the 76th at Kweilin. These three formed the 23rd Fighter Group, the only such unit to ever stand up in the field under combat condi-

tions. They had only fifty-one aircraft, all beat-up P-40s inherited from the A.V.G. (Hill, *"Tex" Hill Flying Tiger*, 177)

The Flying Tigers passed into history, but the war had not. It now fell onto the shoulders of Chennault and the men of the 23rd Fighter Group to continue the struggle.

Part Three

With the US Army Air Forces

14

A New Sheriff in Town

Colonel Robert Scott proved much during his short time in Asia. He showed quiet confidence and was an aggressive, capable fighter pilot. He exhibited great courage and very professional piloting skills. Scott made a point of meeting every man in the AVG, talking to them and learning about the real character of Chennault's group. In doing so, he quietly gained their respect and friendship. Scott considered himself honored to learn combat techniques from them. They were master craftsmen of their trade. He was an eager newcomer. Rank was subservient to experience, as far as Scott was concerned. His modesty and personality won over the aloof, hardened combat veterans.

To Chennault, it seemed a natural fit that Colonel Scott should be chosen as Commander of the 23rd Fighter Group. That was not the way Generals Stilwell and Bissell saw it. They had other people in mind that were more controllable. Chennault, an excellent judge of raw talent, knew better and got his way. Colonel Scott was the right man for the job.

Working under the guidance of Chennault and Scott, units of the 23rd Fighter Group wanted to prove themselves in combat. That was only natural (inbred fighter pilot mentality). By the time the 23rd finished with the Japanese, everyone knew who replaced the Flying Tigers. These men proved themselves worthy of their name-

Figure 14–1. Double Ace Colonel Robert L Scott. Photo probably taken in January 1943, shortly before his return to the US to support fundraising efforts. Back home, he wrote a bestselling novel, *God Is My Co-Pilot,* which was made into a popular 1945 movie. The book detailed how he advanced from a cargo pilot to training as a combat pilot with the Flying Tigers. When the CATF was formed, Chennault chose him as first commanding officer of the 23rd Fighter Group, where he became a double ace. (US Army), www.worldwarphotos.info)

sakes. One need only look at their combat record. These brave new airmen knew most of the privations shared by the 1st AVG and they also fought like Tigers.

Bombers at last!

On July 1st, the 11th Bomb Squadron was activated in China with Colonel Caleb Haynes as commander. Five newly arrived B-25 medium bombers had made the trip over the Hump, supplementing two that arrived earlier. There would have been more, but Bissell had turned the ferry flight into a bombing run over Lashio that caused the loss of four B-25s.

> In the summer of 1942, the CATF was hardly a potent striking force. It had seven B-25 bombers of the 11th Bomb Squadron and forty P-40s to fly against an estimated four hundred Japanese planes based in East China, Indochina, and Burma. Chennault made Colonel Tex Hill leader of the 75th Fighter Squadron and Colonel Ed Rector leader of the 76th, so he had two combat-wise veterans to lead those two squadrons. By August–in spite of the lack of replacements–he was at least ready to hold his own against the Japanese when the monsoons lifted. (Samson, *The Flying Tiger*, 218)

The disbanding of the Flying Tigers was a big deal. Knowledge of the impending dissolution was widespread. Even the Japanese knew exactly when and where this event would happen. In preparation, they made plans to destroy the new arrivals.

> There was now no hiding the fact from the Japanese that the Americans were in East China. The enemy had also known for months that the long-awaited breakup of the Tigers had occurred on July 4th. Enemy pilots had counted the days with eagerness, expecting to reap a glorious harvest of inexperienced P-40 pilots as soon as the AVG bowed out. Just after dawn on the 4th, they opened by sending a dozen I-97 fighters low and fast over Hengyang, ready to start racking up easy victories. (Hill, *"Tex" Hill Flying Tiger*, p. 178)

Chennault anticipated this attack and adjusted accordingly. Thus, his request for Tiger volunteers to stay an additional two weeks. The volunteers made him proud.

> The Tiger veterans wailed away on the shocked Japanese, turning five "Nates" into fireballs before the rest got away to bring the word back to their comrades: whoever the P-40 pilots were now, they must have learned a thing or two before taking over. It was most unwelcome news. (Hill, *"Tex" Hill Flying Tiger*, 178)

Chennault again got the best of Japanese air commanders and pilots. How could such green American pilots fly so well against their seasoned veterans? Just like in the movie *Butch Cassidy and the Sundance Kid*, the stunned Japanese were asking themselves, "Who were those guys?" This surprise would have major negative implications for the pilots of the Rising Sun and their fight for China.

Creating the illusion of many aircraft

Chennault deployed the 75th Fighter Squadron to Hengyang and the 76th to Kweilin. To protect against raids on Kunming or Paoshan from Burma, he dispatched the 16th Fighter Squadron, under Major George Hazlett, to Yunnan-Yi. The 74th Fighter Squadron, under Major Frank Schiel, defended Kunming. The 11th Bomb Squadron's B-25s were based at Kweilin and Hengyang but shuttled back and forth between there and Kunming. Whenever Chennault launched a bombing mission in Indochina, Burma or Thailand, he called the 11th Bomb squadron.

> During the summer of 1942, I became committed to Chennault and through that commitment I became committed to the Chinese nation in its struggle for survival. I took it as a matter of course that the United States would defeat Japan. It was a certainty. Those little bastards could never defeat us. The thought that I might not be around to see it sometimes bothered me a little, but not a helluva lot. (Trumble, *Biography)*

That summer, the best targets were Hong Kong, Canton and Hankow. With escorts going along (just as Chennault preached so fervently in 1931 to his ACTS students), Allied bombers could reach all Japanese targets in China.

Shuttling aircraft between bases, Chennault fooled the Japanese. The High Command in Tokyo believed Allied forces were much larger than they actually were. To further this illusion, the nose cone colors were painted between raids, suggesting to enemy observers that different aircraft were being used. Aircraft numbers were sometimes reversed for the same effect, making Chennault's forces seem larger than they were. This was one of Bedford Forrest's favorite ruses.

> Our tactics were to use every advantage of interior lines of communication, the Chinese warning net, seasonal weather, range of our planes, and the Japanese orthodox combat habits to retain the initiative, jab the Japanese off balance with surprise thrusts at widely separated targets, and keep their numerically superior forces on the defensive. It was the principle of the Confederate cavalry leaders in the War between the States applied to modern air war. With our tiny but mobile air task force, we could cut Japanese communications, destroy supplies, batter their bases, and create confusion in their rear out of all proportion to our admittedly tiny effort. (Chennault, *Way of a Fighter*, 185)

Japan incorrectly assumed the CATF was a major Allied air force deployed against them. Air commanders presumed Chennault's forces were much greater because CATF raids so punished the Japanese.

The Flying Tigers had the ability to launch various strikes over hundreds of miles on the same day. Tokyo incorrectly believed only substantial Allied forces could inflict so much damage. Japan sustained heavy casualties during these lightning raids. Japanese field commanders assumed they were under attack by a complete air wing.

Japanese squadron commanders frequently exaggerated the numbers of P-40s shot down based on pilot claims. Only a major air wing could sustain the claimed 400 CATF airplanes destroyed. Japanese Headquarters received and believed these inaccurate claims as fact. This had an adverse effect on how and where Japan chose to attack the CATF. Had they known, they would have launched sustained attacks until they achieved full destruction.

Know your enemy; then, and only then, attack

Chennault had reliable information about targets he attacked. The Chinese maintained a marvelous underground network of anti-Japanese spies secreted in Burma, Thailand and occupied China. Information from these agents proved invaluable to Chennault, his early warning net and the men of the CATF. Japanese spy networks, although vast with numerous paid agents working in the field, never provided the same quality of information. As a result, on scene commanders proceeded with extreme caution.

> All through July, the Americans attacked targets of opportunity and the Japanese counterattacked. On July 20th, the medium bombers of the 11th Bomb Squadron plastered a cotton yarn factory at Kiukiang, but ten days later, the Japanese sent wave after wave of bombers and fighters over Hengyang in a massive effort to immobilize the advanced air base. The P-40s shot down seventeen of the estimated one hundred and twenty enemy planes that attacked–four of them at night. By the end of July, the claims Chennault had made about destroying the Japanese Air Force did not seem as implausible as earlier. Even the brass in New Delhi began to take notice as the tally rose to twenty-four fighters and twelve bombers destroyed at a cost of five P-40s, one B-25, and no pilot loss. (Samson, *The Flying Tiger*, 220)

If headquarters 2000 miles away doesn't believe your after action report, just shoot down a few more planes. Maybe then they will listen to you. Or even better, they might drop by to see for themselves.

The wisdom of building bases

Chennault, early after his arrival in China in 1937, foresaw the need to build a series of primary and standby air fields across China. Chennault's vision was apparent now. American air forces moved rapidly away from danger. They deployed quickly wherever and wherever they were most needed.

> The dirt and crushed rock fields at Hengyang and Kweilin were being hit regularly by Japanese bombs. Chennault asked the Chinese Army for work battalions and they supplied thousands of coolies for repair brigades. As soon as the Japanese bombed the strips, the coolies would fill up the bomb craters and smooth over the holes. On one day at Kweilin they filled up forty-five craters in two hours. (Samson, *The Flying Tiger*, 223)

Chennault admired the spirit and efficiency of the coolies. He knew with certainty no other people in the world matched the Chinese for their dedication to destroy the Japanese. Japan's spies never understood how the Chinese built and repaired their air fields so quickly. They never discovered any Chinese heavy runway repair equipment. Japanese scouts thought these were hidden quickly, making them impossible to locate. They never dreamed the coolies built these airfields by hand.

> Everything has happened fast in this war, and the organization of the 23rd Fighter Group was no exception. There was no holiday, even if it was activated on the 4th of July. There was no time for celebration. Radio Tokyo started right off with a bang, and we definitely knew hard work was ahead. On the night of July 3rd, Radio Tokyo warned the new American fighter group that they would quickly annihilate them, for it was common knowledge that the experienced AVG personnel were leaving for America. But Tokyo had reckoned without the strategic brain of the General, or the loyalty of those great pilots of the First American Volunteer Group. (Scott, *God is My Copilot*, 157)

Again, brilliant military minds in Japan overlooked the genius of master strategist and tactician Claire Lee Chennault. When the Japanese announced they would come for tea the next day, Chennault made sure there was plenty of hot stuff to go around.

On-the-job training

Although sick with malaria most of that summer, Tex took new squadron pilots into combat. Night missions were launched as required.

Every mission flown by the green pilots added to their knowledge and confidence for combat. Both are vital for making good, effective pursuit pilots. Hill's actions saved many American lives. During this short period, he earned the Distinguished Flying Cross twice.

Hankow

Insistent intelligence of continuing Jap air build up at Hankow pressed heavily on Chennault during the last month of the AVG existence. He sent message after message to the War Department and to Bissell's 10th Air Force in India, almost pleading for fighter planes and pilots to replace the AVG personnel who would leave China en masse after July 4th.

> Only a few fighter planes and Army pilots arrived before July 4th. He asked for volunteers to stay past our official disbandment for fourteen days. Only twenty pilots would agree. Twenty-four ground personnel, including myself, also volunteered to remain. Five of the AVG pilots accepted commissions, five of his staff and twenty-eight ground personnel accepted commissions or enlisted ratings in the Army. The remaining hundred and eighty-seven were adamant in their determination to leave China the soonest possible date. (Trumble, *Biography)*

Chennault directed Hill to fly a single plane night attack against the Japanese base at Hankow. The Japanese planned a sizable operation for early the next morning. Tex dive-bombed the airfield, forcing the Japanese to cancel their mission. Damage done to the airfield prevented the Japanese launch. Hill received the Silver Star for this mission:

> Hill was helped in the defense of Hengyang by such veterans as Ajax Baumler and Johnny Alison, who teamed up to fly night interceptions–shooting down the Japanese bombers in the bright moonlit nights over Hengyang. By the last of August, the Japanese figured the loss ratio from all raids was becoming too high and that Hengyang and Kweilin were not worth the price they were paying in fighters and bombers and the raids slacked off. (Samson, *The Flying Tiger*, 223)

Chennault's intuition did it again. Flying with inferior numbers, he deterred much stronger enemy forces from attacking Chinese cities. Had the Japanese realized what marginal forces faced them, they would have launched a relentless air campaign to destroy the upstart Americans.

> Realizing the danger, the Japs concentrated on wiping out Hengyang and Kweilin, rather than risking raids on Chungking. They lost a lot of planes doing it, but we also lost two wonderful guys, Johnny Petach and "Red" Shamblin. They had volunteered for the extra two weeks. Petach and an AVG nurse, Emma Foster, had been married in Kunming just a short time before. Chennault faced the task of telling her the night of July 10, 1942. (Trumble, *Biography)*

Rescue if Downed

The China Air Task Force was certain about one thing. They could depend on aid from the Chinese peasants. If shot down, forced to bail out or crash land behind Japanese lines, Chennault instructed his men to seek support from local Chinese. Members of the First American Volunteer Group discovered this truth many times while flying in occupied China.

Once war started for the Allies, Chennault had phonetic pamphlets printed, simple area maps drawn and emergency rations issued. In Chinese, he published a card identifying Americans as friends and offered money for their safe return. A patch, called a blood chit, was sewn on their flight jacket and contained the same information.

Many peasants or guerrilla forces rescued downed airmen. Successful rescue brought tremendous honor and respect for the Chinese that took downed airmen through Japanese lines. Chinese civilians gave aid to injured or lost airmen, knowing full well the Japanese executed them if caught.

Rules of engagement were set by the conquering Japanese forces. Brutality, butchery and torture were the orders of the day for Japanese soldiers. These "brave" sons of Nippon gave no quarter to military or civilian captives. They were often tortured or executed for the smallest offenses. The Allies, especially guerilla bands, responded in kind. War in China was hard fought and merciless. Quoting Air Marshal Sir Arthur Harris in a speech to Bomber Command personnel in 1942, "They had sown the wind and now they would reap the whirlwind."

This was a fight to the finish. No quarter was given to anyone! Assistance to downed Allied airmen is highlighted by the rescue of P-51 pilot Jack Gadberry. He was shot down, presumed dead by his squadron mates. Don Lopez recalled:

> One evening, four or five other pilots and I were in one of the hostel rooms–feasting on boned chicken and pineapple slices in lieu of the mess hall food–when suddenly someone walked into the dimly lit room. He was dressed all in black and appeared to be Chinese, except for his size. As he came into the light, we were happy to see that it was Jack, one of the four pilots shot down over Hengyang on November 11. We had heard nothing of him since then and assumed that he had been killed or captured. (Lopez, *Teeth of Tiger*, 203)

When someone was killed, his possessions were usually gathered, sorted and returned to his family. Items like pistols, knives or other weapons, deemed unsuitable for return, were auctioned off to his surviving friends. Money gained was forwarded on to his family. Food or canned goods were divided among his friends to be eaten. In China, preserved food was so scarce it became a very valuable commodity.

> Momentarily, it was a bit embarrassing; we were just finishing off the last two cans of his private stock of food brought from India. His roommates had been eating it regularly for about a month, since we had given up hope of his returning. (Lopez, *Teeth of Tiger*, 203)

Lopez and his partners in crime spent the evening listening to harrowing tales of Gadbury's bailout and escape. He was lucky to avoid capture, crediting this to the rapid Chinese response and Jack's fortitude. His P-51 was on fire and before he knew it, the enemy attacked. Forced to bail out over the airfield at Hengyang, by good fortune, his parachute landed half a mile from the field.

Watchful Chinese peasants saw him and quickly led him to safety. A short distance away was a farmyard. They knew of an underground oven with a heavy metal lid. They hid him inside. It minimally accommodated his body, even in the fetal position. For three days, he remained in the oven until the Japanese gave up.

> Several times, as best as he could tell from the sounds, Japanese on horseback and on foot were within a few feet of his hideout. They finally abandoned the search, and the Chinese were able to get him out at night and turn him over to the guerrillas, who led him through Japanese territory by a long and tortuous route with many layovers. He had heard from the guerrillas that one of the pilots on his mission had gone down with the plane, but he did not know who it was. After the war, we learned that Miller had been killed and Taylor had been taken prisoner. (Lopez, *Teeth of Tiger*, 203–204)

Without the tremendous courage and daring of Chinese peasants and guerrilla forces operating behind Japanese lines, there would have been far fewer escapes. Knowing the risks these Chinese heroes took every day, American airmen felt a strong sense of obligation to fight their mutual enemy. Throughout the long war, brave Chinese civilians and paramilitary groups rescued hundreds of Allied airmen. Knowing the reprisals against them and their families, they continued aiding Allied pilots. It is with an eternal sense of gratitude that these airmen remember the Chinese people.

Reinforcements

Reinforcements promised by Bissell for June finally arrived in late September. Six additional B-25s with a bunch of worn-out P-40s supplemented Chennault's tiny force. It wasn't much, but it was just enough. Chennault and his flyers always got the last, the least of anything they needed. Most of the men flying under Chennault believed this was deliberate. Chennault agreed with them.

In September 1942, Chennault received his greatest resource in the China campaign, the arrival of twenty fully trained fighter pilots. All were well versed in long range navigation, gunnery, dive-bombing and tight formation flying. These were the traits Chennault requested in 1940 when searching for recruits for the AVG. Most of these men volunteered but were refused because they were defending the vital Panama Canal. Late or not, Chennault happily accepted them. Panama veterans produced many of the finest pilots that served in China. Chennault's combat reports frequently cited these men for bravery and excellent performance.

> For more than a year, they were the backbone of the China fighter squadrons. After they arrived, the C.A.T.F. smacked less of a primary training school and more of a combat group. The Japanese soon felt the difference. (Chennault, *Way of a Fighter*, 193)

Changing group commanders

Commander Scott was not only a skilled leader but also a driven fighter pilot, even though he was older and higher ranked than usual. He named his P-40 "Old Exterminator" and frequently took it on solo missions. By September, he became an ace, and an object of attention for a number of war correspondents with his pithy quotes.

This was one occasion the Air Corps made no mistake. Any great organization needs great leadership, great personnel and a can-do attitude. These key elements in place, adverse conditions are more easily handled. China in 1942 was one of those challenging places, and Chennault and Colonel Scott were the right two commanders.

Colonel Scott was Chennault's group leader and trusted friend. Scott could not understand why there was never enough of anything to keep his group fighting. Frustrated, Scott went to see Chennault at headquarters. He wanted to understand why this problem continued.

> The Old Man looked at me as he must have looked at hundreds of impetuous people before me who just knew part of the facts and little or none of the intrigue. He did not tell me to get out and STOP worrying him. Instead, he walked over to the map of sprawling China. There among the faded greens and oranges he pointed out the winding trail...Believe it or not, there are Chinese out there, working and dying too. And they've been waiting a lot longer than you. You fly your P-40 with two hundred gallons of aviation fuel that lasts two and a half hours for five hundred miles if you have a little full-throttle work to do. When you come back, you fill up with another two hundred gallons. Fifty four-gallon tins that came all the way across the 681 miles of the Burma Road, or four fifty-five gallon drums from over the Hump. You know how that got here, but do you know how it goes on another five hundred miles to where we can use it to fight?...Sometimes when you are on your way east and you cross this road–if you can call it a road–look down and you'll see what I mean...There'll be coolies rolling fifty-five gallon drums. Sometimes they've rolled them all the way from Kunming to Chanyi, almost a hundred miles...You see, Scott, the Chinese have to do it by hand mostly. (Scott, *Flying Tiger*, 187)

Scott said he was beginning to understand. Later he said:

> Of all the honors that I ever have received or ever will receive, the greatest to me will always be the honor of being given command of that great group, under the command of General Claire L. Chennault. (Scott, *God is my Copilot*, 156)

By December, Scott became a double ace. He enjoyed the resulting publicity and became quite famous back in the US, which had an unintended consequence. In early January 1943, Washington ordered him to return to the US to support fundraising efforts! Before Scott

FIGURE 14–2. COLONEL CASEY VINCENT. Vincent replaced Scott in 1943. Here, he is flying Scott's P-40, #7 *Old Exterminator*, over the lake near Kunming. Note the prominent landslide scar on the mountain; all pilots in the theater knew this landmark as Scarface. (US ARMY), 103602AC)

was recalled, Bissell assigned Colonel Casey Vincent to Chennault's staff, assuming he would put some spit-and-polish into the CATF.

> Vincent liked and respected Chennault but had nevertheless concluded that any similarity between the CATF and a military organization was purely accidental. (Byrd, *Chennault*, 162)

Chennault chose Vincent to replace Scott as group commander. Continuing Scott's tradition, Casey was soon flying combat missions to learn how it was done from the experts.

15

The Green-Headed Monster

With a serious air campaign beginning, jealousy and discontent reared its ugly head. Tension between the 10th Air Force and the China Air Task Force increased daily. With Chennault's increasing success, resentment increased steadily at Bissell's Headquarters.

Bissell deliberately sent only minimal supplies to China and Chennault. Using supply as a weapon, Bissell controlled Chennault, getting him to kowtow through the supply chain. Stilwell and Bissell controlled destination and cargo tonnage inside the 10th Air Force sphere of operations. They took every advantage of that leverage, limiting supplies sent to China.

> In April 1942, the First Ferrying Group under the U.S. Tenth Air Force began flying supplies from India over the mountains to China to support Chiang Kai-shek's armies–a trickle at first, and a torrent later. Between Assam [in northern India] and China are the eastern Himalayas, with great spurs of parallel mountain ranges extending north and south at altitudes of up to twenty-two thousand feet. To cross these mountains was popularly known as "crossing the Hump"—a distance of about five hundred miles. The Hump route, over these towering, snow-covered mountains, was often in the range of Japanese airfields with their deadly Zeroes. (Spencer, *Flying the Hump*, 11–12)

Bissell ensured there were never enough supplies at 10th Air Force to allow increased tonnage for China. Stilwell believed victory in Asia required huge numbers of American ground troops. He stockpiled tons of supplies for his exclusive use. These were necessary to fight the conventional land campaign Stilwell and Bissell anticipated. Refusing to accept reality, Stilwell never understood America would not invest the forces required. Believing his plan would eventually gain acceptance, Stilwell and Bissell refused all requests made by Chiang and Chennault. Fearing this would drain Stilwell's necessary resources, only minimal aid was forthcoming for any China operation.

This was a stupid, selfish system complicated by a deliberately confusing chain of command over two thousand miles from Chennault. Stupidity remained in vogue as long as Stilwell and Bissell remained.

> A writer for Life magazine described Chennault as "being obsessed with one consuming passion, to beat the Japanese....[this] has established this wrinkled, scar faced, half deaf, fifty-one-year-old, ex-barnstorming pilot as the one genius that the war on the Asiatic mainland has produced." (Spencer, *Flying the Hump*, 16–17)

Bissell's operation was at best obtuse and inefficient. At worst, it was deliberate and criminal. His counterproductive, vindictive and silly organization eventually cost the Allies all the East China airfields, supplies and very effective observation network.

Chennault's only war time goals were to defeat the Japanese and aid China. He didn't care who got the credit for success. Chennault later wrote:

> I always felt that Bissell prized a snappy salute from a perfectly uniformed staff officer more than a Japanese plane shot down in flames. As Bissell's subordinate, I obeyed his orders to the letter, but I never developed much respect for his ability, either as a military administrator or tactician. (Chennault, *Way of a Fighter*, 201)

Inbuilt redundancy produced friction and failure for operations conducted by the 10th Air Force in China. Jealousy, envy and prejudice directed against Chennault bear responsibility for deaths occurring within that theater. In military jargon, "responsibility is borne by the commander." In this case responsibility goes directly to Generals Stilwell and Bissell.

> During that winter the radio between Delhi and Kunming cracked steadily with an exchange of terse, bitter "eyes alone" messages between Bissell and

> myself. Many of these exchanges were over apparent trivia. But maintaining morale of the Americans who were fighting in China with inadequate equipment and short supply of everything involved much apparent trivia. Our morale problem during long periods of inactivity forced by lack of supplies was terrific. I never ceased to be amazed at how pilots and ground crews survived these intervals of bitter boredom and always managed to produce so well in the pinches. (Chennault, *Way of a Fighter*, 201)

Forced deployments are tough on a military man; combat deployments overseas are tougher still. Deployed forces need care and must be supplied with basic things. Soap, warm clothes, razor blades, cigarettes, promotions and decorations for combat valor make a man feel wanted. These are simple things to provide but create contention when not.

Tom Cotton wrote to his grandmother after arriving in China,

> The mail situation is very bad with none coming in. China is a rather dirty place. I should say this part of China is dirty. The majority of the people live in poverty that we never see in the States. America's poorest people are well to do in comparison. Disease is rampant among the Chinese, making things worse. We have Chinese pilots in our squadron now. They are just as intelligent and educated as we are. They come from the Jap occupied territory and have been trained in the States. (Cornelius and Short, *Ding Hao*, 273)

When a soldier feels overlooked, his mind sees things from a different perspective. If not well fed, well cared for, or made to feel unappreciated, serious problems soon follow. Particularly in combat morale and discipline crumble. You no longer have a command, you have a mob. Mob mentality is not a good thing in combat, nothing positive ever follows. Chennault in Kunming knew this. Bissell, sitting in luxury and comfort on Per Diem Hill in India did not. This was a recipe for disaster. Soon enough, Generals Bissell and Stilwell learned this truth the hard way.

> To understand the value of the American effort in China and the need for the Hump airlift, one must recall that although China had been fighting the Japanese for ten years, the country was still not prepared for modern war. Chinese soldiers were poorly trained and equipped and had little will to fight. Chiang Kai-shek's armies were rotten with graft, politics, and greed. Chinese generals, selected on the basis of political influence, often refused to lead their armies into combat. Chinese pilots who "washed out" in flight training were certified as proficient so their families would not "lose face"... At one time, in 1941, China had fewer than seventy-five planes that were operational, and Japan was able to march with almost no resistance. (Spencer, *Flying the Hump*, 14–15)

I wonder to whom Bissell and Stilwell thought that rolling head would belong?

Bissell informed Chennault that he (Bissell), sitting in Delhi, India, was a better judge of a CATF fighter pilot's condition in China. He dismissed the on-scene commander's recommendation. Bissell forced AVG veterans Tex Hill and Ed Rector to stay in China for an extra month. Both contracted malaria and dysentery.

These pilots continuously flew combat for a year. This was a primary example of "one assmanship" (military term for higher headquarters stupidity) displayed by the 10th Air Force command. Bissell deliberately withheld available supplies marked for Chennault and his men. Bissell was not the most popular man in China, at least not among Chennault's flyers. CATF fighter pilots sat in their airplanes or alert shacks, waiting long hours between flights. Many spent their time searching for elaborate X-rated insults directed at 10th Air Force staff officers. These fighting men ranked him somewhere between cow poo and the bottom of one's shoe. They told each other they were off to the latrine to take a Bissell!

> At Kunming they taught the Chinese coolies who carried cargo and passengers' bags from the transports to greet the new arrivals with what can be politely translated as "nuts to Bissell." This, the fighter boys assured the coolies, was a standard American greeting. The coolies, pleased with their knowledge of English, religiously greeted each transport with a resounding chorus of "nuts to Bissell." This was a great sport for the browned-off fighter boys until one day when General Bissell himself stepped off a transport at Kunming and was greeted by grinning coolies shouting "nuts to Bissell." He was not amused. (Chennault, *Way of a Fighter*, 202-203)

You know Chennault smiled enough to unbend his perpetual scowl when this happened. Facing death every day, it's good to have something that brings a smile to your face.

Chennault's Chinese eyes and ears

Chennault realized early on that accurate information was critical. One of his first actions in China was the creation of an effective early warning system to detect and report all Japanese activities. His reporting network employed many Chinese spies. Many operated behind Japanese lines. Their natural animosity toward the Japanese was motivation enough. Chinese volunteered in the thousands. Almost a million men served in Chennault's warning network. Chennault always received correct information. These men, at great personal danger, obtained pertinent, timely information on Japanese intentions.

Throughout China, hundreds of observation and spy cells reported vital information on a daily basis. The volume of timely, accurate information was sorted, prioritized and delivered to Chennault constantly and provided the means to insure victory over the Japanese.

> One of our tasks was to direct air strikes against the Japanese via radio. A communication set had been installed in our Jeep, which we hid in bamboo growth overlooking the enemy lines. Electricity for the radio transmitter came from a hand-cranked generator either of us two agents had to crank wildly, a job more wearing than the spy mission itself! The enemy knew very well that someone was directing our American planes to their hiding places but was unable to figure out from where the guidance came since we constantly moved our Jeep. (Neumann, *Herman the German*, 117)

Missions undertaken by Chennault's observers provided much vital information for follow-on attacks. Correct intelligence was one of the key elements for the success of Chennault's aviators.

> Boats and ferries loaded with Japanese troops, light tanks and cavalry were our favorite bombing targets. We suggested the number of our aircraft to use and what typeof bomb to drop, and gave precise directions when to leave the home base to catch the Japanese in the midst of their river crossing...By following our directions from the ground, they knew precisely when and where to drop antipersonnel or explosive bombs even if our aircrews could not see the targets from above. (Neumann, *Herman the German*, 117)

During the grim period, this was the only visible evidence of American aid or offensive spirit in China. Millions of Chinese needed their courage and determination to continue the fight. Allied fighting reached its lowest ebb in China. They desperately sought ways to resist Japanese aggression. What was needed most from America was encouragement and a supply of weapons. The 14th Air Force was the smallest numbered American air force ever commanded by a general officer. For sure, it was the most raggedy. With the 14th Air Force, Chennault would get another star.

> Its paperwork was poor, and salutes were scarce, but when the signals were called for combat, it never missed a play. (Chennault, *Way of a Fighter*, 206)

Every gallon of aviation gasoline was worth its weight in gold. General Bissell foolishly ordered Chennault fly leaflet drops over enemy territory. His intention was to dishearten Japanese soldiers and encourage them to give up and go home. Disgruntled by what he perceived as a wasteful effort, Chennault did what any old soldier would do. He put aside personal feelings and said yes sir. Good soldiers obey orders.A leaflet to the Chinese in Hong Kong said:

> Greetings from your Friends. This leaflet is dropped by aircraft of the United Nations. Every day now more and more planes will fly over Japanese-occupied countries, wreaking vengeance on the Japanese invaders. We are getting stronger every day. Already the Japs are retreating in New Guinea, the Solomons and the South Seas. We are destroying their planes and sinking their ships. Five of their aircraft carriers are at the bottom of the sea. The Allied Air Force in China gets more deadly every day. The United Nations will never forget the suffering you have endured, the rape of your womenfolk, the loss of your homes, the confiscation of your goods, the humiliation forced on you by the Japanese beasts. But keep your spirits up! Side by side with your brothers of the Chinese Army, we shall come back to Hong Kong. (Neumann, *Herman the German*, 112)

Messages of this type were of limited value but were the only source of information for those trapped in Japanese captivity. While this was a nice gesture, it would have been more effective if accompanied by planes bombing and strafing Japanese positions. Bissell never learned. (If I were a Japanese soldier, a pamphlet would not get me to go home, but a bomb certainly would.) A little more thought can go a very long way.

Put me in, Coach, I'm ready to play

With expansion of American involvement in China, problems of supply multiplied. More men meant more supplies required, more aircraft meant more fuel, oil, ammunition, more spare parts, more, more, more of everything. No one at 10th Air Force Headquarters perceived any kind of supply difficulty. Chennault and the CATF were just fussing, like always. This seemed logical to Stilwell and Bissell, whose supplies were stockpiled to the roof in Delhi.

No one else complained of shortages, so what was the problem? The two top men at 10th Air Force felt Chennault was only complaining again. While the AVG was in action, Stilwell got more requests from the annoying Chennault than anyone else. Stilwell considered him a constant complainer. Failure to meet unrealistic supply quotas promised to Chennault became a sticky problem. As it turned out, Chennault had a well-placed friend in Washington. Maybe he could help.

> In September both Stilwell and Bissell approved a plan for increasing the C.A.T.F. to a strength of 105 fighters, 12 bombers, and 4 photo planes, and providing replacements to keep it at strength. They also agreed to provide the C.A.T.F. with a monthly tonnage 1,968 tons delivered by air over the Hump from India. They never even came close to fulfilling any of these firm commitments. By January of 1943 the C.A.T.F. was getting only 300 tons a month over the Hump. In March, out of a total allocation of 1,000 tons, only 615 were delivered. At the beginning of each month a collection of Tenth Air Force and C.B.I. Theater officers solemnly drew up a schedule of guaranteed Hump deliveries. Just as regularly at the end of the month, C.A.T.F. records showed receipt of 30 to 50 percent of the supplies promised. (Chennault, *Way of a Fighter*, 303)

10th Air Force shortages might be because of fluctuation in the war effort in other places, or it could be sloppy bookkeeping at the supply section. Who knows. One thing was certain: all units run by Stilwell and Bissell maintained an excellent administration section. Both prided themselves on paperwork and efficiency. Many officers serving under them believed proper record keeping was more important to them than combat efficiency. Because of pettiness and jealousy, Bissell and Stilwell deliberately sabotaged Chennault's efforts in China.

> As early as August 13, 1942, I tried to impress on Stilwell the urgent need for action in China by the following radio: "Situation in Orient appears extremely critical STOP Unrest India STOP Jap threat to our transport service from Burma STOP Determined opposition to MacArthur's offensive in islands and Chinese inactivity due to our failure to deliver requested air support and lend-lease supplies all contain elements gravest danger to American war effort STOP Axis continues victorious on all fronts while Allies have failed to launch successful offensive anywhere STOP Action of a small effective US air force in China would do much to relieve that situation STOP It could destroy much of the war materials flowing through and around Formosa to the southern islands for use against MacArthur STOP Inspire Chinese ground forces to action against Japanese-occupied areas STOP Neutralize Jap air efforts in Burma and IndoChina STOP Relieve immediate Jap threat

> to India STOP Safeguard our air transportation line to China and supply a successful offensive to inspire all Allied powers STOP. (Chennault, *Way of a Fighter*, 210–211)

10th Air Force headquarters thought Chennault made exaggerated claims about achieving victory with minimum resources. Conventional military wisdom scoffed that these claims were unrealistic, impractical, and unattainable. Receipt of the requested quota Chennault claimed would have provided the material necessary for victory. The ring knockers at 10th Air Force never gave him the opportunity.

> I have demonstrated practicability employment of force for more than two months with a handful of planes STOP If total of 500 planes bombers and fighters plus 100 transports given me plus complete authority in this theater I will accept full responsibility for attainment of objective listed STOP Nowhere else in the world can one US plane destroy from 8 to 10 enemy planes while damaging so much other enemy operations at the same time STOP To do this I require number of planes mentioned above delivered in increments echeloned over five months STOP Due to extreme gravity situation immediate action necessary STOP Urge this radio be forwarded Generals Marshall and Arnold for decision and necessary action. (Chennault, *Way of a Fighter*, 210–211)

Stilwell, Bissell, and 10th Air Force personnel never attempted meeting Chennault's requested minimum requirements. This reeks of complicity at the highest levels. Headquarters pundits laughed at Chennault's unrealistic requests. Promises made by President Roosevelt were never delivered to Chennault. Stilwell provided a much reduced tonnage than promised.

> The year that Stilwell and Bissell allowed the Hump potential to remain undeveloped [summer 1942 to end of Trident Conference in May 1943] was a tremendous loss both to the Chinese war effort and to our entire Pacific strategy. (Chennault, *Way of a Fighter*, 235)

The 10th Air Force had no intention of supplying the tonnage requested by Chennault. Petty jealousy eventually cost the Allies all of East China, lock, stock and barrel. This was no way to run a war, and it was becoming more certain by the day which general was right and which was wrong.

The theater commander was the only person with power to reduce tonnage deployed in his area. Reduced tonnage must receive support from the air corps commander, Bissell. Collusion from General Arnold, who hand-delivered the President's message ordering increased tonnage for China, never gave that promise to Chennault. A Presidential order was deliberately disobeyed. These three senior officers deliberately refused to honor Roosvelt's directions. This action, committed during wartime, is inconceivable. Many military men would call that a treasonous act. This was and still is a stain on their military careers.

> This early failure of the Hump was not due primarily to the technical problems of the aerial supply line but to the contempt of Bissell and Stilwell for this method of supply. Stilwell's ignorance of the airlift potential was understandable but why Bissell, an airman, persisted in labeling the Hump operation as impractical I was never able to comprehend. Bissell stoutly maintained that to deliver 5,000 tons a month to Kunming from India required a fleet of 300 transports and 25 airfields. Less than a year after his fantastic estimate the Air Transport Command was delivering 10,000 tons per month to Kunming using 150 transports and half a dozen fields. (Chennault, *Way of a Fighter*, 203)

Bissell, and possibly Stilwell, for personal reasons, wanted Chennault to fail in China. Restriction of his supply projections assured no victory for him. Both men did everything possible to make Chennault lose face (a very serious concept for the Chinese). By addressing petty personal problems, a significant number of men died. Stubbornness and stupidity (although all too common) are not good qualities for command. Deliberate refusal to honor supply quotas resulted in loss of much of Eastern China. Chennault's airfields and magnificent early warning net were gone. Chennault wished Stilwell and Bissell were gone too!

Neither Stilwell nor Bissell were incapable commanders. There must be some other reason for their failure. Mountains of needed resources surrounded them. That this was a deliberate action to spite the commander of operations in China is a real possibility. Whether that commander was Chiang Kai-shek or Chennault is not clear, but the results remained the same.

Stilwell and Bissell tarnished their reputations, military careers and helped destroy Nationalist China. Chennault's reputation in China is pristine. This service is considered his best. History has a funny way of biting one in the behind when least expected. People in China still regard Chennault as their savior. Nuts to Bissell!

Chennault and his American forces in China in 1943 knew as long as they were under the 10th Air Force, they would never receive adequate supply. Remaining under the direct control of Generals Stilwell and Bissell guaranteed nothing would ever be done. The CATF did its best anyway.

End of the CATF

During nine months, the CATF destroyed 149 enemy planes in the air and 85 more probable kills while losing 16 P-40s. In 65 bombing missions, Japanese fighters only penetrated CATF fighter cover once. A single B-25 was shot down by the Japanese. CATF dropped 314 tons of bombs, but those bombs shattered Japanese security. Most importantly, they validated Chennault's theory. A small air force operating on a semi-starvation diet of supplies could hold its own against larger Japanese forces. Case closed.

Apparently, someone in Washington felt the same way.

On March 10, 1943, a radio message from Washington announced the formation of the 14th Air Force. Operating from China, it would be under Chennault's command (no more Bissell!). With that short message, the CATF passed into history. Even with its grim beginning and shortage of supplies, the China Air Task Force left a proud combat record.

16

Major General Chennault

Now commander of the newly created 14th Air Force, Major General Chennault was in position to challenge theater commander Lieutenant General Stilwell about the conduct of the war in the CBI.

> The basic issue between us was whether the war should be fought in Burma on the well defended Japanese perimeter where the enemy was prepared and anxious to fight as part of his over-all defensive strategy or whether it should be fought in China, where immediate blows could be struck at the enemy's vitals and where they were neither prepared nor willing to fight. (Chennault, *Way of a Fighter*, 220)

The defining area of conflict between Chennault and Stilwell revolved around which war should be waged. Stilwell favored a massive ground assault. Chennault wanted an air campaign. Stilwell's plan required large numbers of American troops, substantial supply and considerable time to plan. A ground offensive involved direct assault through dense jungle against well-trained fighters in fortified positions. The cost in men and material expended made this method prohibitively expensive.

Chennault wanted an air campaign using resources already in place, supplied and ready to attack immediately. An air offensive did not have to penetrate jungle terrain, mountains with defensive forti-

Figure 16–1. Major General Chennault. Promoted on March 10, 1943, he now headed the Fourteenth Air Force and no longer reported to General Bissell. (US Army, 49424AC)

fications, large investments of men, and could begin immediately. It looked like an obvious choice.

The most damaging aspects of Stilwell's proposal were massive troop deployment, excessive training of Chinese soldiers and time required before the assault could begin. He planned to lead this land campaign to recapture all of Burma. In doing so, he would remove the stain on his military record (like Douglas MacArthur in the Philippine Islands). It was a needless campaign. In its broadest scheme, Stilwell's offensive also included operations against the Japanese Home Islands.

This was sound military planning as far as it went. An optimistic estimate of its earliest completion was ten years hence. Stilwell never once mentioned the word airplane in his battle plan. He was a ground officer, and that was all he knew.

> His supply plan was based on mules and trucks operating over roads that could not supply a single American division. He was content to fight a strictly ground war with his beloved "men in the trenches." (Chennault, *Way of a Fighter*, 221)

These tactics mimicked 1917 warfare, but mobility on the battlefield had changed over the first three years of this war. Mobility and speed now ruled. Stilwell knew nothing of these changes. Warfare's new darlings, location change and rapid deployment, were not utilized in his operational planning. Tactics changed, but Stilwell didn't. He was reluctant to accept this new concept of warfare.

It is advisable, even prudent, to select commanders well versed in new tactics of modern warfare they are expected to fight. From 1939 until 1941, the German Wehrmacht were masters of mobility on the battlefield. Success followed their every move.

Stilwell's experience as a 1900's infantry commander ensured he knew nothing of the capabilities and uses of aviation. Chennault called him the best four star company commander in the US Army (not a compliment). His strategy was sound during the Hundred Years War or the Somme Offensive of the First World War. Casualties of the same proportions were expected. Stilwell wanted to fight an infantry war in an air war theater. Mountains, rivers, and prepared enemy defensive lines are detrimental to ground warfare, but not of much use against airplanes. Stilwell was an old time warrior in a new time war.

> Stilwell's indifference to intelligence is also well documented. Neither in Burma nor in China did his C.B.I. headquarters make any effort to organize an effective intelligence network. Lacking any genuine intelligence necessary for a sound military appraisal of enemy intentions and capabilities it is not surprising that Stilwell was often wrong about both. Stilwell not only lacked his own sources of intelligence but was also prone to discount any intelligence offered him from other quarters. (Chennault, *Way of a Fighter*, 312)

Stilwell was set in his ways and refused to adapt his strategy or methods. His operations required head- on assault into fortified Japanese positions. (similar to the Marine Corps. doctrine) He believed in the "4F" military solution, (the mixed company version goes like this) you must "find them, fix them, flank them and finish them. Guess which F word was changed to protect the innocent? An effective but casualty causing strategy. (Remember General Grant's old maxim, "he who gets there firstest with the mostest carries the day.")

Regrettably, the Japanese Army opposing Stilwell was brave, dedicated, and adaptable. Masters of fortified defensive techniques welcomed head-on attacks. General Stilwell was brave beyond description on the battlefield. Once an able soldier, the new art of mobile war passed him by. People in Washington were beginning to think the same thing. This war in Asia needed perhaps guidance in a different direction.

> My plan was basically the same broad concept submitted to Wendell Willkie of using China as a platform for mounting an air offensive against the vitals of Japan. Aircraft strength requirements were boosted to 150 fighters, 70 medium bombers, and 35 heavy bombers. The Hump tonnage required was 4,790 tons for the first three months, building up to a monthly total of 7,129 tons delivered to the Fourteenth Air Force in China thereafter. (Chennault, *Way of a Fighter*, 221)

Stilwell and his minions would have none of this air campaign nonsense as directed by Chennault from his distant bases in China. This was a case of penny wise and pound foolish. Stilwell's plan again bears the heavy thumbprint of a certain General Bissell. Allowing petty jealousies to interfere with sound tactical judgment cost the Allies a precious opportunity to do serious harm to the Japanese. Score another one for the ring knockers. Had Stilwell gotten support for his ground offensive, it was by then too expensive for serious consideration. Stilwell was beating a dead horse. If the Allies were to achieve victory in Asia, they must use a new plan.

> Philosophical differences between Stilwell and Chennault caused much personal animosity and rancor for the rest of the conflict. Cynthia Chennault confirmed there was only hostility and disrespect between the two men during the war and afterwards. (Cynthia Chennault, *conversation with author*, 2018)

No common ground existed between these two. There was no chance for a positive working relationship. In boxing terms, Stilwell fought as a heavyweight cautiously moving forward to deliver a massive blow. Each aspect is carefully thought out and cautiously positioned. If the blow connects, the fight is finished. Planning, timing and execution are required for this type of victory.

Chennault fought like a bantamweight, always moving, darting in and out, avoiding stunning blows. Parry and spar, strike when any weakness is observed. Sting the opponent, annoy him and make him bleed. Repeat the blow as often as possible and wear him down. Go in fast, strike the blow and retreat out of reach. Exasperate the opponent, tire him out, make him react foolishly, then strike him down.

Both styles are appropriate when correctly fitted to circumstances. Both styles can produce victory. In this instance, what was needed was a correct assessment. Which method worked best with the resources on hand? (In the words of a great boxer, it was necessary to "float like a butterfly and sting like a bee!") Warfare in Asia was fluid because of the great distances involved. Neither man recognized the strategic importance of China was slowly shifting.

Orders? What orders?

Chennault and his pilots worked under impossible conditions. Chances for success were reduced not so much by Japanese in the field but by American commanders at 10th Air Force in India.

> On March 3, 1943, I received word of my promotion to major general. Seven days later, the Fourteenth Air Force was activated in China, and my command was separated from that of Bissell's. In mid-March a four engine Liberator B-24 bomber group arrived in India for assignment to the Fourteenth. On March 27, Brigadier General Edgar E."Buzz" Glenn arrived to become my chief of staff. He brought the first sizable contingent of staff officers, the first to reach China since the war began. (Chennault, *Way of a Fighter*, 216)

General Glenn also brought orders from Hap Arnold to Stilwell. Per President Roosevelt's request, control of Hump transport operations was transferred from Bissell to Chennault immediately. The President refrained from issuing an executive order on the transfer only after General Arnold's personal assurance was given. This action was ill-advised (never delivered) and virtually doomed all Sino-American coordination for the rest of the war.

> This was the best news of all since bitter experience had proved that a field commander who had no control over his supply lines is in a dangerous predicament. Glenn transmitted these orders in writing to Stilwell, who, as was his wont when unpleasant facts were forcibly called to his attention, promptly filed them and forgot. Stilwell never executed the President's order. The Hump route remained under Bissell. (Chennault, *Way of a Fighter,* 216–217)

Observe the "old boy" network at its worst here. Senior officers deliberately disobeyed a direct order from a sitting President in wartime. It was done specifically to prevent a rival officer from achieving success in the field. Some people might call this treason. (I certainly would.) If Stilwell was in the Chinese Army, his head would be missing.

Rank has its privileges

Not everything happening to Chennault at this time was bad. Many changes for the better were in the pipeline. Every military man knows rank has its privileges. This was true for Chennault, now a Major General and commander of the 14th Air Force.

> My personal household consisted of my chief of staff, Brigadier General "Buzz" Glenn; Colonel Tom Gentry, my personal physician and chief flight surgeon of the Fourteenth; and Captain Joe Alsop, who functioned as mess officer. Two Chinese houseboys, Riley and Gunboat; a cook; my Chinese driver Wong, who had been with me since 1938; and Joe the Dachshund, completed the staff. (Chennault, *Way of a Fighter*, 228)

Together, this tiny crew made a good team and provided a sense of home, family and comfort for Chennault throughout his remaining war years. These people gave support and trust for the "Old Man" when he needed it most. His friends and staff were available when he wanted to hunt, fish, talk or have a congenial drink. His men considered selection for any of these tasks an honor.

Time alone with Chennault was special. His most trusted companion was a little dog known alternately as Joe, Joe Dog, or Joe Dash. Chennault carried his pet no matter where he went or who was with him. In this innocuous way, he could show kindness and his soft side without detriment to his warrior image. Joe Dog was of great comfort to Chennault when he most needed it. Many times Joe was seen sitting on the table while Chennault ate a hurried meal. Joe was there to distract him from the many concerns of command, and that was what he did. Joe the dog had an interesting story and Billy McDonald II loved telling it.

> Just before Chennault and I left for Kunming, I found a man with some puppies. I bought one and gave it to Colonel Chennault. He became fond of this puppy and they were constant companions. Joe slept at the foot of the Colonel's bed and the Colonel taught him numerous tricks–sitting up, barking on command. Joe also became a great hunting dog. (McDonald and Evenson, *The Shadow Tiger*, p. 145)

Billy McDonald III remembers a story of his father's about Joe Dog. One time while on a hunting trip with several friends:

> Chennault shot a sandhill crane and his little dog Joe jumped into the cold water to retrieve it. Unfortunately for the dog, the crane was only wounded. As Joe dog splashed away, the large bird got the best of him. Before Joe went down for the third time, Chennault stripped off all of his outer garments and jumped into the ice cold water to save the foundering little dog. Cold, wet and exhausted, Chennault retrieved the wounded bird, the little dog and himself. On shore and freezing, all the participants [except the crane] survived the experience. This just shows that you can take the boy out of the country, but you can't take the country out of the boy. (Billy McDonald III, *conversation with author*, 2018)

Being away from home and family for so many years had a heavy effect on Chennault. Letters from Nell were less frequent. It was obvious his family situation was not working. With each passing month, the gap between Chennault and his family (especially with Nell) widened. Time and distance slowly changed family dynamics and nothing could be done about it.

Now it was to his surrogate family and closest friends Chennault turned to for comfort and comradery. He relied on them. They provided friendship, trusted counsel, and relaxation as necessary. They gave him hope for the immediate future, provided hunting partners and encouraged physical activities (such as baseball, his favorite sport).

These provided a few moments of precious time free from stress. At night, there was always a chess competition. Chennault even took these games too seriously. As often as prudent, he would sit with the little dog sitting on his lap while sipping Bourbon whisky with a red pepper in it. He claimed it mellowed out the strong taste of the liquor. Chennault was never known to drink to excess during the entire war.

Like most men at war, Chennault continued sending letters home, most often to his girls. They were still at an age he worried about their wellbeing.

All six Chennault boys served in the war effort. Jack flew in the frigid skies of the Aleutians (a very cold, windy and nasty place), Max did air traffic control at Morrison Field where he kept trying to get a posting in combat. Son Pat was a Mustang (P-51) pilot in England. Charles was an AAF radio mechanic. Bobby was an aviation cadet earning his wings and could not wait to get started. David was in the Pacific serving aboard the cruiser USS *Helena*.

Even rank and high command are no protection for family members at war. Chennault worried and prayed like any other parent, but he was proud of his boys. You can be both afraid and proud at the same time!

The Hump

Japanese gains in Southeast Asia, Burma and China isolated the Nationalists from any outside source of resupply. The Allies were hard pressed with island campaigns being fought elsewhere in the Pacific. If China needed food, munitions, troops and all the necessities for waging war, she could not get them from the outside world. Japan was strangling China just like a big Anaconda snake.

The only possible supply route since the loss of Burma and the famous Burma Road was by air from India to China. At best, it was a very thin lifeline. Flights took off from Assam, India to Kunming, China. This path flew over treacherous, impassable jungle, high mountains and deep river crossings. To reach India, it was necessary to cross the intimidating Himalaya Mountain Range, aptly named the rooftop of the world.

Here exists some of the worst weather on Earth. Violent frontal systems, extreme windshear and cumulus clouds can reduce visibility to nothing in a few minutes' time. Severe mountain turbulence can tear an airplane in flight to pieces within moments.

Making matters worse, there were no mechanical aids to navigation, no detailed maps of the area, and mountain tops frequently exceeded the capabilities of transport aircraft (C-46 and C-47) flying the route. The first thing newly assigned pilots learned on arrival was don't fly into clouds unless you have no choice. Those that listened might survive.

Keep in mind these airplanes were unarmed and flying with inexperienced crews all the while crossing across Japanese occupied Burma. There were very few boring days for Hump pilots. Each flight provided many challenges for the aircrew. Life expectancy was only a matter of months. Air Transport Command pilots flying this route called it the Aluminum Highway. Navigators are not necessary–just follow the crashed C-47s. Odds were fifty-fifty of dying on your tour as not. Those aren't even good odds in poker.

All the new pilots pondered their fate. High mountains above and dense jungle below were all bad choices for airmen in distress. Few crashed pilots ever returned to base. This was a formula for disaster and shouldn't have been allowed. Green, unskilled and overly brave young American pilots trying to conquer the greatest moun tain range in the world. Flying through Japanese airspace just added to the excitement. Terrible, unforecast weather didn't help much either. Better be sure your G.I. insurance was current and paid up.

Colonel Robert Scott remembered his earlier days flying the hump, before he transformed himself into a P-40 driver and fighter pilot:

> I was stationed at the tea plantation of Shallote in Dinjan, eastern Assam, at India's far eastern tip. I was executive officer, as well as operations chief, of the Assam-Burma-China Ferry Command, an organization which was all of a week old. It was spring 1942, and I had a "stable" of about ten DC-3's and C-47's. They were unarmed old Douglas transports. (Scott, *Flying Tiger*, 115)

The transports were too slow to get away from any fighter and could not even climb high enough to clear the taller mountains along their route. Avoiding them required navigation, but the Japanese held most of the route and that ruled out radio navigation fixes. They flew by "dead reckoning": you fly this fast that way for this long, and you should be here (as long as you guessed the wind drift correctly). Flying these aircraft was like playing Russian roulette with four rounds in six chambers. It was possible to survive, but the odds were stacked against you.

> With them we were supposed to cross the highest mountains in the world every day, flying over the Hump, the toughest air route there was. We'd take off from Dinjan in the big loop of the Brahmaputra and climb across the Naga Hills on a course of 119 degrees for Kunming, China, 500 miles away. Most of the way we barely topped the clutching mountains. There would be black monsoon clouds veiling the crags of the Himalayas and reaching down to overflow the steamy jungles. (Scott, *Flying Tiger*, 115)

In the brief intervals of clear weather, when pilots could see what lay below, they found themselves wishing for clouds to materialize and blot out the hazardous peaks that loomed at their altitude. Their theory was if you couldn't see them, they couldn't hurt you.

> We called these clouds "hard centered cumulus" but never let them daunt us. Whenever I would mention them to Cooper, he'd yell at me, "Want to live forever, Scotty?" smiling with the raw joke. But I knew he really wasn't joking. "You wanted ***out*** of the training command, remember? You didn't appreciate that safe, comfortable, reliable desk. (Scott, *Flying Tiger*, 115–116)

No wonder this was the most dangerous flying job of the war. What is amazing is not how many pilots died flying the "Hump" but rather how so many survived. By the end of the war, flying this dangerous route claimed 585 air transport planes and thousands of personnel. These brave men flew against terrible odds and paid a terrible price. Their contribution to victory in China is incalculable.

Most pilots assigned to the Air Transport Command were young second lieutenants fresh out of pilot training, with only one or two hundred hours of flying time. Not much of that time was on instruments and mostly flown over Texas. There is not much bad weather during summer in Texas and hardly any 20,000 foot mountains to run into. These young men were combat pilots. They wanted to fight for their country. All volunteered for pilot training, went where they were sent, and did what they were supposed to do. During their assignment, a great number of them died.

> Throughout the operations over the Hump, the jungles of Burma, miles below their aircraft, held a special fear for the airmen. To go down was almost certain death-to try to walk out was almost impossible. The dark, silent jungle was so thickly matted with undergrowth, it was impossible to penetrate except

> on trails by natives. And headhunters lived in these jungles. (Spencer, *Flying the Hump*, 26)

It took tremendous courage to do what Hump pilots did. The United States owes these brave men a great debt. By the time ATC aircrews served their tours over the Hump there were very few old, bold pilots! Surviving the Hump was one of the great accomplishments these brave, inexperienced flyers ever achieved.

> During its three years of Hump operations, A.T.C. delivered 736,374 tons to China and lost 468 transports —nearly as many planes as the Fourteenth lost during three years of combat. To A.T.C. Colonels Edward Alexander and Thomas Hardin and General Earl Hoag goes the credit for building up the Hump during the critical period when prospects for success seemed dim. Bill Tunner took over toward the end and pushed the tonnage to incredible figures. (Chennault, *Way of a Fighter*, 234)

These quantities were in line with Chennault's earlier requests that were laughed down by Stilwell and Bissell. Here again, we catch a whiff of his genius.

China National Aviation Corporation (CNAC)

A commercial airline, CNAC, first began flying supplies over the Hump. CNAC pilots made the first survey flight in November 1941. They wanted to prove it could be done, in case all routes to the east were closed. They flew regularly from Kunming to Rangoon, and on to Calcutta. When the Japanese took Hong Kong, Hanoi, and Burma, going over the Hump was the only way to get to India.

During the war years, CNAC crossed the Hump with consistency, fewer losses and greater efficiency than ATC troop carriers or combat cargo squadrons. It was safe enough for experienced personnel of commercial caliber. Flying better aircraft with commercially trained (more experienced) airline quality pilots enhanced chances for survival.

These civilian pilots also got a lot more money for doing the same job. That's war for you. Civilians take chances and get paid a fortune—military pilots take the same chances and die.

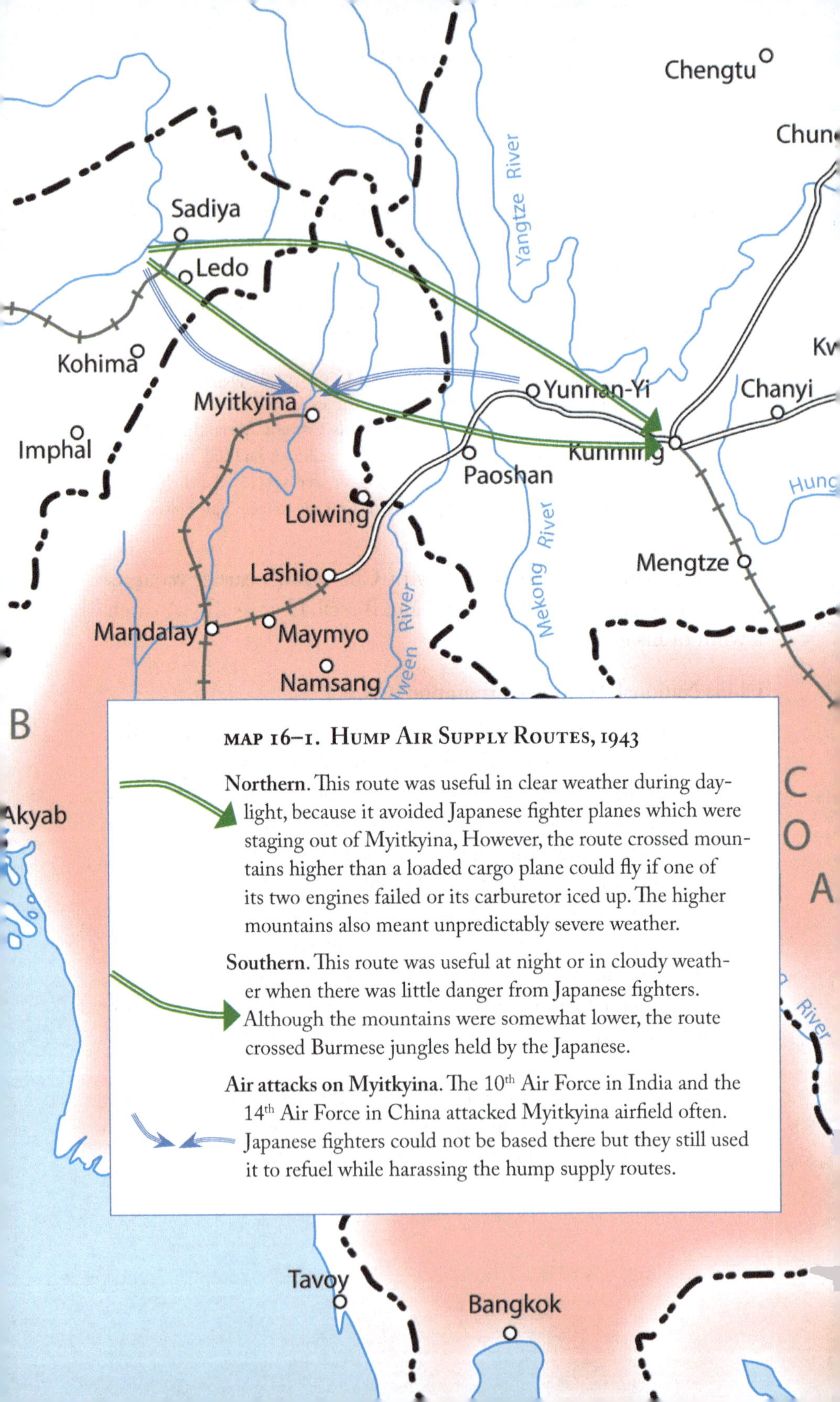

MAP 16–1. HUMP AIR SUPPLY ROUTES, 1943

Northern. This route was useful in clear weather during daylight, because it avoided Japanese fighter planes which were staging out of Myitkyina, However, the route crossed mountains higher than a loaded cargo plane could fly if one of its two engines failed or its carburetor iced up. The higher mountains also meant unpredictably severe weather.

Southern. This route was useful at night or in cloudy weather when there was little danger from Japanese fighters. Although the mountains were somewhat lower, the route crossed Burmese jungles held by the Japanese.

Air attacks on Myitkyina. The 10th Air Force in India and the 14th Air Force in China attacked Myitkyina airfield often. Japanese fighters could not be based there but they still used it to refuel while harassing the hump supply routes.

Organization

Positive things were now happening for Chennault and the 14th Air Force. For the first time since he came to China in 1937, supplies arrived in sufficient quantities where and when they were needed. Aircraft and pilots arrived, as well as the heavy B-24 bombers Chennault wanted. New fighter types came also, twin-engined P-38s, the rugged P-47s, and the nimble long-range P-51s.

The organization changed with the influx. New composite wings combined bombers and fighters for more effective coordination. Willing and qualified personnel made things happen. Men new to the service did not come with pre-existing prejudices. Vast increases in manpower did not provide enough time to develop new ring knockers. Preconceived ideas of what a military professional should be were not part of any officer's ninety-day training. These men

FIGURE 16–2. UNLOADING CARGO FROM A CNAC C-47. Chinese workers unload boxes of 20mm ball ammunition at Suifu [Ipin], China, on December 11, 1944. Former AVG pilot Bill Bartling [back to camera in sweater] dickers with the Chinese gas-man as to how much fuel to keep for the return trip. Chinese mechanics fix a flat on the truck in the background. (G. BULL, 160-32)

brought the qualities of eagerness, flexibility and an overwhelming desire to get the job done.

New guys brought a new philosophy to the conflict. All they wanted was "to win the war, get back home and get on with the rest of their lives." Their favorite saying was "home alive in '45" and they would do anything to make that happen. No one said, "It can't be done." Now they asked, "What does it take to get the job done?" Positive attitudes go a long way in war time.

Formosa Raid

With new personnel, supplies and aircraft in place, Chennault launched a bold series of attacks against the Japanese. One of the most successful was a long-distance raid on the main Japanese airfield on the island of Formosa (Taiwan).

> Next morning, very early on Thanksgiving Day 1943, our P-40's left our temporary airfield, heading north to fool anyone who might possibly report to the Japanese intelligence the goings-on at Suichwan. Thirty minutes later, all P-51's, P-38's and B-25's took off in tight formation in a southeasterly direction, toward the Formosa Strait...It was imperative the most direct route be flown to conserve every bit of fuel-and to be able to return to Suichwan...... What would our planes find at their "Target Formosa?" (Neumann, *Herman the German*, 115)

In any combat area, complacency is a dangerous habit. This is just what happened here. The Japanese thought they were safe from air attack and allowed their defensive posture to become lax. All it takes to overcome false assumptions is an out of the box thinker. Chennault was an out of the box thinker. The Japanese were soon paying for their laziness and lack of vision.

> That's why the Japanese used "safe" Formosa, with its excellent runways and complete aircraft service facilities, as a staging point for planes to be ferried from Japan to islands in the South Pacific. Our fighters and bombers were carefully scheduled so as not to interfere with each other: They had to arrive over the target in a tight sequence, at precisely the right moment. Shortage of fuel did not permit any mix up. True to the preflight briefing, our pilots found Japanese fighters and bombers lined up, wing tip to wing tip, in double rows, on both sides of the runways. (Neuman, *Herman the German*, 115)

Tight Integration for Fighters and Bombers

Chennault created composite wings to encourage coordinated efforts, rather than competition. Moving these Wings as necessary, the 14th Air Force was able to play both defense and offense, attacking and pushing back against Japanese incursions.

68th Composite Wing
- 23rd Fighter Group (P-40, P-51)
- 308th Bomb Group (B-24 Heavy Bomber)

69th Composite Wing
- 51st Fighter Group (P-40, P-38, P-51)
- 341st Bomb Group (B-25 Medium Bomber)

Chinese-American Composite Wing
- 3rd Fighter Group (P-40, P-51)
- 5th Fighter Group (P-40, P-51)
- 1st Bomb Group (B-25 Medium Bomber)

312nd Fighter Wing
- 81st Fighter Group (P-40, P-47)
- 33rd Fighter Group (P-38, P-47)
- 311st Fighter Group (P-38, P-51)

Other
- 426th & 427th Night Fighter Squadrons (P-61)
- 341st Bomb Group (B-25 Medium Bomber)
- 443rd Troop Carrier Group (C-47)

The Formosa Raid reflected new thinking, new confidence, and new men. The operation was brilliantly conceived by Chennault and executed flawlessly by 14th Air Force pilots. The mission was declared a major success. The attack caused great consternation to the Japanese militarily and their sense of security was shattered. If their previously un-hit sanctuary could be attacked, they were no longer safe anywhere.

If the Japanese based large numbers of aircraft, naval vessels, or troop concentrations in an area, that place would never be secure again.

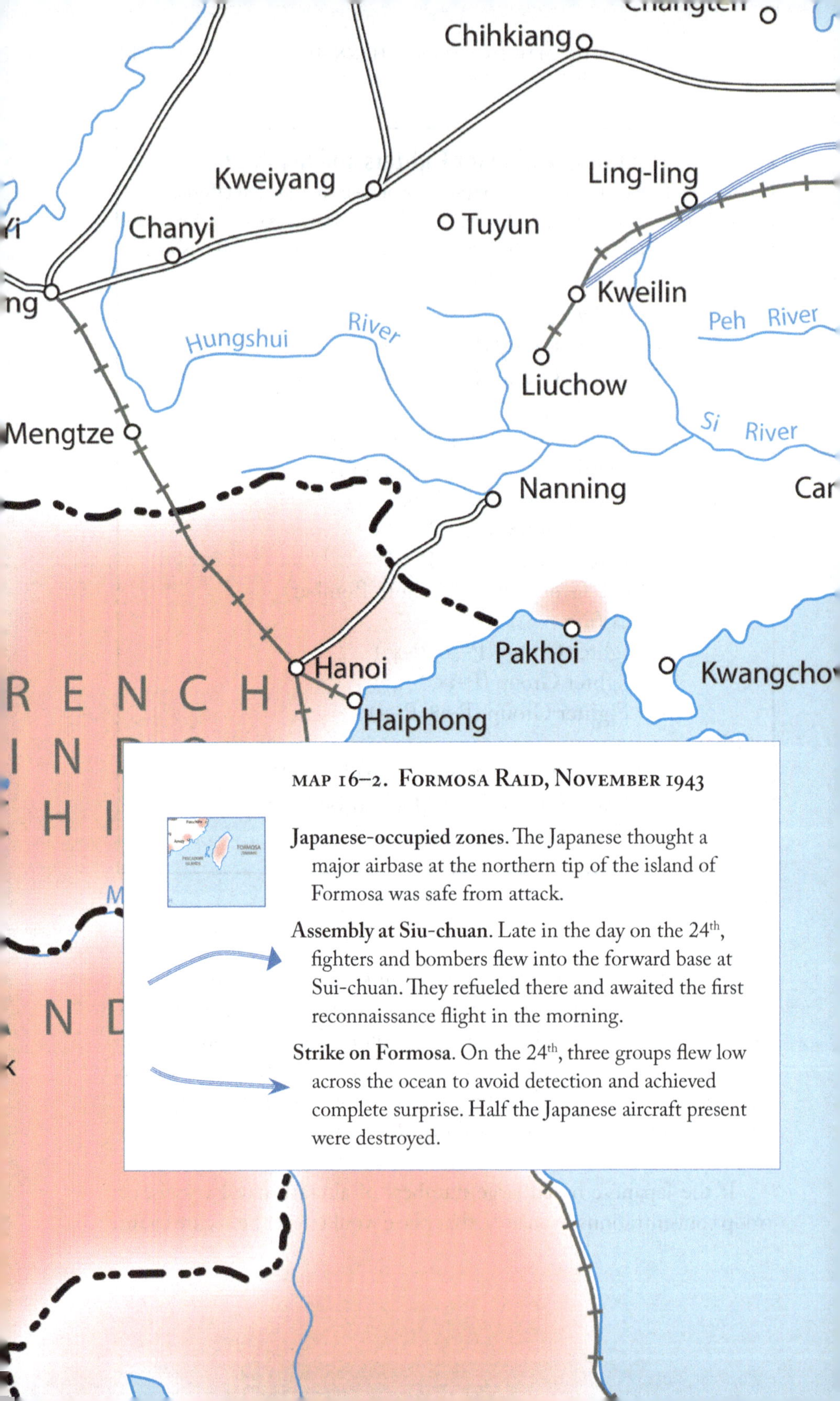

MAP 16–2. FORMOSA RAID, NOVEMBER 1943

Japanese-occupied zones. The Japanese thought a major airbase at the northern tip of the island of Formosa was safe from attack.

Assembly at Siu-chuan. Late in the day on the 24th, fighters and bombers flew into the forward base at Sui-chuan. They refueled there and awaited the first reconnaissance flight in the morning.

Strike on Formosa. On the 24th, three groups flew low across the ocean to avoid detection and achieved complete surprise. Half the Japanese aircraft present were destroyed.

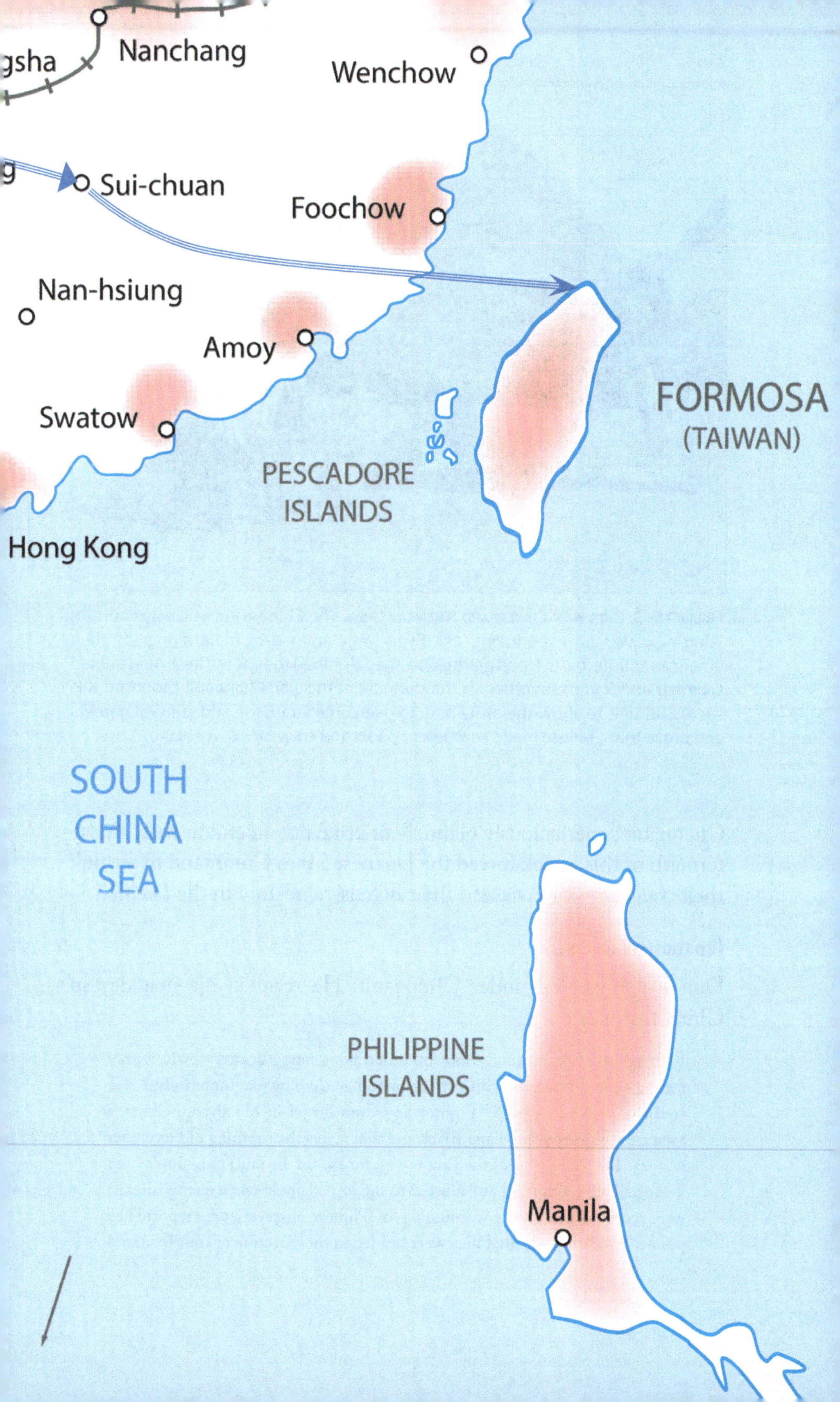

Nanchang
Wenchow
Sui-chuan
Foochow
Nan-hsiung
Amoy
Swatow
Hong Kong
PESCADORE ISLANDS
FORMOSA (TAIWAN)
SOUTH CHINA SEA
PHILIPPINE ISLANDS
Manila

Figure 16–3. B-25 with Chinese and American Crew. The Chinese-American Composite Wing (CACW) was an experiment. CAF flyers were trained in India alongside their American counterparts. Upon graduation, they all joined the CACW. These men trained together and fought together for the duration of the war. Chennault proposed this successful unit to show the army how to overcome language and cultural barriers and prove that Chinese made competent pilots and crew. (US Army, 93082AC)

Chennault's meticulously planned airstrike put an end to that. The aftermath of this attack forced the Japanese High Command to rethink their war plans for Asia and their defensive posture in the Pacific.

Japanese response

Don Lopez served under Chennault. He recalled his first day in China this way:

> I struggled awake to the clamor of a loud gong being beaten slowly, with a feeling of unfamiliarity with my surroundings. As I awoke more fully, I realized that I was in the 75th Fighter Squadron hostel in Hengyang, China, a forward air base on the Siang River, and that it was the morning of December 12, 1943. The clanging was the Jing boa [Chinese for air raid] gong indicating a one ball alert. One ball was hoisted to the top of a pole when enemy aircraft were reported aloft by the warning net of Chinese observers, who reported by phone. Two balls indicated they were headed in the direction of your base, and

Figure 16–4. B-24 Heavy Bombers with P-51 Escorts. Both of these aircraft had a longer range than the P-40s, opening up a wider range of targets to the 14th Air Force. American technology had delivered both bombers and fighters that outclassed anything in the Japanese arsenal. With its 8-hour endurance, an unloaded B-24 could also range all across the South China Sea doing reconnaissance to locate Japanese shipping for attack by B-25s and submarines. (US Army, 52924AC)

> three balls meant that fighters should scramble to intercept the enemy. In addition to the balls, a gong was sounded slowly for one ball, faster for two and incessantly for three. (Lopez, *Teeth of Tiger*, 13)

Alert is always rush, rush, rush, then stand down. The enemy changes course and is not attacking. Then it is a matter of waiting, waiting and waiting. A pilot's life is hours of boredom followed by minutes of sheer terror. That's the way it goes. Your stomach never knows quite what condition it should be in. Generally somewhere between throwing up and tied in knots.

> Suddenly the radio came to life and the voice of Major Richardson crackled in my earphones. White squadron, scramble! Able and Baker flights stay low to intercept the bombers. Charlie Flight climb to intercept the fighters. On receiving a nod from Jim Anning, I immediately put my starter switch to en-

> ergize position, gave the engine a few shots of prime, turned on the ignition, and engaged the starter. The engine caught right away, and I put the mixture control into auto-rich and adjusted the throttle. To me, there is still no more exciting sound than the whine of an inertia starter, followed by the hesitant then deafening roar of a powerful in-line engine. (Lopez, *Teeth of Tiger*, 3)

Aggressive targeting

Bombing was disheartening to infantry or cavalry as well. Chennault declared anything moved by ground, air or sea were primary targets. This air campaign was so effective the Japanese restricted troop movements to nighttime or in daytime during bad weather only.

Attacking Japanese shipping off the coast of China became a speciality. First, long range B-24 bombers conducted extensive sweeps of the seas, covering a huge area in one mission. Location of any Japanese shipping went straight to US Navy submarines, and the B-25 bomb groups. Chennault's men added additional 50-caliber guns to their aircraft and used low-level skip-bombing tactics that proved deadly to Japanese shipping.

Wiltz Segura and rescue by Chinese

The 75th Squadron, which included Charlie Olsen, Tom Cotton and Wiltz Segura, went to Changsha on a strafing mission called for by Chennault. Launching early in the morning, Segura later recalled this, his first combat.

> We were just past the city called Changsha by Tungting Lake. There were supposed to be some troops crossing the lake, hauling supplies in small boats. It was a strafing mission, and we used eight airplanes, four on the bottom and four at the top. That gave us top cover in case the Japs would intercept and hold them until we could climb up and get away. I was on the bottom. When we shot up all of our ammunition, we always saved a hundred rounds and then we'd go and fly top cover. The Japs were in small boats on the lake. There was no place for them to hide. The troops were paddling like mad, and we came down upon them in a line abreast. I picked up a boat, a sampan ahead of me. Must have been six or seven troops in it with baggage. I looked at them through my sight, coming down waiting to get in range. Just as I pulled the trigger, I realized I'd never shot a man before. Well, all of a sudden I thought, "That's the enemy. He's trying to destroy my way of life. It's like a ball game. Let him have it." I split the boat in half, saw people jump on both sides. I blew up everything. After that, it was one after another for a year and a half. It got

Figure 16–5. B-25 The Green Hornet. An American mechanic and a Chinese sentry examine a 50-caliber machine gun. Note the additional heavy machine guns mounted along the sides of the fuselage. In front of each gun, the white spots show the effect of the muzzle blast. Extra armor plates were added to protect the fuselage. Photo taken at Kweilin, August 2, 1944. (US Army, 69168AC)

> very impersonal after that. We were mad at them for what they were doing with our prisoners in the Philippines. (Cornelius and Short, *Ding Hao*, 350)[1]

Chinese peasants and irregular units constantly rescued downed Allied airmen. Americans were brought through Japanese lines every day. Even when crews were shot down over water, Chinese were there to pick them up. Near Hong Kong and Hainan Island or any-

1 Author's Note: Now this is a funny thing. Wiltz Segura was from New Iberia, Louisiana, the small town where I grew up. As a boy, I knew and respected the General. He was a fearsome-looking man, and I always thought it wise to keep a safe distance. I knew he flew during the war and was an ace. That was all. Flying for Chennault in China was a surprise. Imagine my shock when I came across his exploits while researching this book. (*I wish now I knew him better).*

where at sea, junks appeared out of nowhere ready to fish them out. Hurry up before Japanese patrol boats arrive.

Airmen eventually ended up back at their bases. Sometimes it took months to get them home. Most were thin and exhausted from long treks through the Chinese interior. Even when it took months of walking, these airmen returned safely and were grateful. Like Dorothy in the Wizard of Oz exclaimed, "There's no place like home, Toto."

Early in the war, Chennault asked the Chinese to make silk flags, "blood Chits," he called them. Worn on the backs of their leather jackets, the chits stated that the wearer was a friend of China, and to please help.

> The flags were written in Chinese and sometimes in Burmese. American fliers were never turned away. The Chinese knew these men were fighting for China. (Chennault, *p. 217)*

Wiltz Segura endured this memorable experience twice. But for Chennault's training and the courage of the Chinese people, he believed he would not have survived the war. Segura told his brother Perry,

> Chennault enforced a demanding initiation ritual for all us incoming pilots. Each man was told about guerilla activities going on in his sector. Chennault insisted Intel-officers briefed us pilots where help could be found if we got shot down. Without those guys, lots of pilots would have been lost. (Perry Segura, *conversation with author*, 2018)

With fighter pilots in a scramble, everything goes crazy. Rush to the airplane quick as you can. Strap in or not, start the engine, now taxi like a madman. Turn to see as much as possible to the front. With that big Allison in-line engine, there is no forward visibility at all. Get into position, hurry and take off. Whatever you do, don't get strafed on the ground. Wiltz Segura remembered,

> I could not see, the windshield was icy, and on top of the ice there was dust. I stuck my head out of the cockpit watching the P-40 in front of me. Then I saw the whole plane disappear in front of my eyes. He had taxied right into a bomb crater. I stopped abruptly and waited for another plane to come by and then followed him. And sure enough, he disappeared too. I stopped again

> until another came by and, thus working my way through the bomb craters, came to the end of the field. Must have lost three leaders getting there. I'm not even strapped in when I finally get to the end of the runway and someone said, "Get off the ground!" (Cornelius and Short, *Ding Hao*, 369)

Sometimes it even defies death just to get to the takeoff point, much less into the air. At least there were very few boring moments once you got your engine started.

Stilwell recalled

Stilwell focussed on regaining Burma and creating the Ledo Road to connect India and China. His Chinese troops fought well and his plan was proceeding in north Burma, when the Japanese Army tried to invade India from central Burma. The British under General Slim were able to contain this drive at Imphal and Kohima. Slim embraced airpower, both in direct interdiction, in aerial resupply, and in troop mobility, eventually soundly defeating the Japanese Army. This eased the way for Stilwell in Burma. However, Stilwell had become more and more adversarial with the Chinese and Chiang.

Chennault complied with President Roosevelt's request to submit information directly to him via personal letter. Such out of channel communications were normal for Roosevelt, but frequently caused problems for their authors. In this case, they hurt Chennault when higher headquarters (Ring knockers) learned of his out-of-system communication. Senior officers disapproved of these types of correspondence. They felt this was disloyal, disrespectful, and talking out of school.

Higher headquarters personnel (Stilwell and the ring knockers) would not acknowledge any successful operations conducted by Chennault. They did not want him to gain further recognition from the 14th Air Force and its commander. One significant report brought matters to a head.

> In this report, Chennault maintained adamantly that Japan could be defeated from China, that she could be defeated by a properly equipped and supplied air force so small as to be considered ridiculous in other theaters of the war. He did not mention or criticize either Stilwell or Bissell by name, but he did imply that he must be granted positive direct command. He cited his success

> with the organization and the leadership of the First American Volunteer Group. He stressed the importance of keeping the Chinese as friends of America at any cost and indicated that this vital need was as much for the future fight against communism as for the war with Japan. He said the military task was a relatively simple one which had so far been complicated by an unwieldy and illogical military organization under the command of men who understood neither aerial warfare nor the internal problems of China. (Scott, *Flying Tiger*, 215)

The report went forward to the highest levels of government. It said a mouthful, both about the present war and those that would follow. How sad for America and China that this communication was quickly forgotten and overlooked.

> On October 19, Stilwell was relieved of his command by the War Department and the C.B.I. Theater was split into two sections. Major General Albert Coady Wedemeyer left Lord Louis Mountbatten's staff to head the new China Theater and was soon given the third star of a lieutenant general. Stilwell's former chief of staff, Major General Daniel Isom Sultan, commanded the new India-Burma Theater. I was not sorry to see him go. This wry, wiry little man carried with him out of China the curious mixture of pride, prejudice, misunderstanding, and misdirected determination that poisoned Sino-American relations for two critical years and hamstrung effective Allied strategy on the Asiatic continent. (Chennault, *Way of a Fighter*, 309)

The wrong men in the right place at the right time did more damage to Chinese-American relations in a few months than the rest of the twentieth century could repair. *Quelle dommage!* (too bad!).

> Stilwell departed, leaving a legacy that is still costing both the United States and China heavily. During his regime he bled China white and squandered most of the American resources available in the C.B.I. Theater on a military campaign of minor value, while the fundamental problem of preserving China as a base of military operations and a strong postwar ally was subverted to his personal quest for supreme command. Ironically, if Stilwell had attained his goal of supreme command, it would have almost certainly produced little change in the China situation, because he had never developed an effective technique for dealing with the Chinese. The fact that he would have been authorized to shout soldier's orders instead of threats over lend-lease would have made no fundamental difference in the results. (Chennault, *Way of a Fighter*, 322)

Behind him, Stilwell left a dreadful legacy. It included an entrenched belief that all Chinese leadership was worthless. This evaluation unfortunately was held by many senior American military and political leaders. Most, including Stilwell, thought Chinese troops would not fight to aid the Allied cause nor would they ever do so. Trust and respect were never given to the Chinese. Stilwell believed Chiang only wanted lend-lease supplies for use after this war concluded. He had no respect for Chiang or Madame as leaders of free China.

These facts and many others damaged Sino-American relations during the remainder of this war and afterwards. They were not the credentials Americans should bring to China. And in a time of war, no less. In terrible times, for everyone involved, sometimes a mistake in judgment or a misspoken word overheard causes unintended pain or disruption. These insults, deliberate or not, will take a long time to repair, possibly never.

Chennault heard of Stilwell's dismissal with a sense of relief, glad for his departure. Free from Stilwell and his staff, he hoped for a better working relationship with the new man in charge.

Chennault gets a new boss

General Albert Coady Wedemeyer replaced Stilwell on October 27, 1944. In addition to being in charge of all US forces in China, he was named a Chief of Staff to Chiang. Chennault said of his new boss,

> Wedemeyer is a man of great personal integrity and studied fairness in his dealings with others. He found it possible to deal effectively with the Generalissimo and other Chinese leaders on a frank yet dignified basis without rising to truculence or sinking to subservience. As a result he found his advice accepted, his plans carried out, and his opinions valued. Without the formal command over the Chinese armies that Stilwell always claimed was necessary for any action, Wedemeyer and his American assistants whipped twenty Chinese divisions into top shape, equipped them with American arms and supplies, and created the nucleus of a truly modern Chinese army. This was all done primarily with supplies that were air-lifted over the Hump, since the Stilwell Road provided little except trucks. (Chennault, *Way of a Fighter*, 323)

New soldiers with new attitudes can make all the difference during difficult times. No more can't do it. Now all that was heard was how we can best do it to get the job done. The war was changing, Chennault could see it. Perhaps if he looked a little closer, he would have seen his fate too.

These were the happy times for Chennault and his allies. The start of operations of the newly formed 14th Air Force was filled with pilots straining to get off of their leash. With a growing stock of P-40s and a few P-51s, Chennault was ready to get at the Japanese. On arrival, these new aircraft were sent to the 23rd Fighter group. Their pilots were ready to come on-stage and, using a 14th Air Force technical term, "kick-ass." With Chennault's training regimen, he personally ensured this would happen. This was a unique and remarkable air campaign in World War II.

> Fighter pilots lived a precarious existence in such a remote sector of the war. Intense air action over enemy territory alternated with long interludes of boredom and inactivity. Life was austere, with poor food, bouts of dysentery, rat infested barracks, and irregular mail deliveries. Heavily laden C-47's and C-46's kept the Fourteenth Air Force going by flying supplies over the Hump [the Himalayas] from India. It took six gallons of fuel to deliver one precious gallon of gasoline to the beleaguered American units. (Samson, *Teeth of Tiger*, vii)

What other countries could fight under such circumstances and in such a desolate area of the world? It took all the ingenuity and supply capacity the United States could muster, but when those supplies arrived, they went where most needed. This was the miracle of a modern day airlift. Stilwell never grasped this concept. He neither knew how nor when to use it. Chennault's airlift was hard pressed to deliver, but it never failed. Using every aircraft under his command, including the mighty B-24, supplies, munitions and fuel arrived in ever-increasing tonnage. This speaks volumes about the courage and determination of all those men involved. Well Done!

> Countless sorties were flown over the rugged mountainous terrain of China, strafing enemy supply lines and attacking troop concentrations. These bold air-operations-perhaps the most distant projection of American air power in Asia at the time-did not go unchallenged: the Japanese launched numerous air raids and a major ground offensive to destroy the American air contingent.

> Even as the Japanese advanced, American pilots scored many notable victories, a paradox of victory in retreat. (Samson, *Teeth of Tiger*, vii)

Warfare in reverse

In what proved the strangest combat of the Second World War, Chennault conducted an aggressive, successful air campaign while constantly retreating. He inflicted far greater casualties against advancing Japanese forces than the retreating Allies incurred. Here was the new (air) cavalry at its best. Mimicking the tactics of Bedford Forrest or Jeb Stuart, Chennault launched lightning raids against the attacking Japanese army. His campaign consisted of strafing attacks and aerial interdiction by small numbers of aircraft. Used against massed hordes of Japanese ground troops, these raids inflicted terrible destruction.

Chennault, a masterful strategist, exposed the vulnerability of Japan's ground forces. Without friendly air cover, Japanese columns were defenseless. It was almost too easy. Chennault's well planned strikes delivered at widely different locations brought death and destruction from above. The 14th Air Force attacks were so effective the Japanese only moved their troop columns at night or in bad weather. Chennault ensured whenever and wherever the Japanese went, they got a tremendous beating.

17

The 14th Air Force

The 14th Air Force raid on Formosa from the East China Bases on Thanksgiving Day in 1943 shocked the Japanese planners, but it hadn't been repeated. However, Chennault's anti-shipping campaign in the Formosa Straits and the South China Seas was being run from those same bases. Ships en route to Japan were being sunk on a daily basis.

So the Japanese planners began to consider a major invasion in central China aimed at seizing control of the entire railway from Peking to Canton, and connecting into Indochina. This would provide a way of moving supplies from South-East Asia to Japan without being vulnerable to American submarines or Chennault's aircraft, which would be driven from their advanced bases in East China. It might be possible to continue north from there to Chungking and knock China out of the war.

Operation Ichi-Go

The Japanese plan required a massive three step operation with punishing battles of attrition during each phase. The provinces of Henan, Hunan and Guangxi were where the Chinese army would be bled white. After China sustained such a humiliating slaughter of her troops, she must surrender. No nation could sustain such casualties and continue fighting.

Phase one, Operation Kogo: Attack and capture central Henan in April-May to open the railroad between Peking and Hankow. Phase two, Operation Togo 1: Drive south from Hankow towards Changsha and Hengyang in June-July. Destroy the enemy, capture both cities, inflict overwhelming casualties against the Chinese. Phase three, Operations Togo 2 and Togo 3: Follow the railroads south to capture the airbases at Lingling, Kweilin and Liuzhou. Force them to commit all available reserves. Inflict maximum casualties. Attack with ruthless aggression, decimate the Chinese and destroy their will to fight. Deny American air forces the ability to attack Japanese shipping in the China Sea, Formosa Sea or South China Sea. Without Chennault's eastern airfields, Japan itself was beyond the range of their bombers. Continue to the Indochina border and complete the rail link.

Ichi-Go was the largest operation ever launched by Japan in China. 500,000 well-trained men with modern equipment and supporting air forces were committed to battle. China opposed these units with 1,000,000 men and Chennault's 14th Air Force. The stage was now set. Ichi-Go (meaning the final throw of the dice) massed as many Japanese troops as the initial attacking force in 1937. If nothing else, the operation validated the failure of Japan's strategy for quick victory. It also exposed significant flaws in military planning for Asian victory. The Japanese High Command consistently miscalculated the time requirements necessary for total victory in China.

Defense of the east China bases

From activation of the 14th Air Force until his relief of command, Chennault engaged enemy ground troops using aircraft only. In military history this was never attempted before. Genius or resourcefulness did not cause this to happen, necessity did. The defense of east China demanded infantry against infantry. Holding territory requires boots on the ground. Without ground opposition, the Japanese had free access to movement. The difficulty for Chennault was there was no infantry to be had.

Whatever fighting was employed, it must come from the air. By this time in the war Chiang Kai-shek was reluctant to commit his vast ground armies against the Japanese. Unless China received ad-

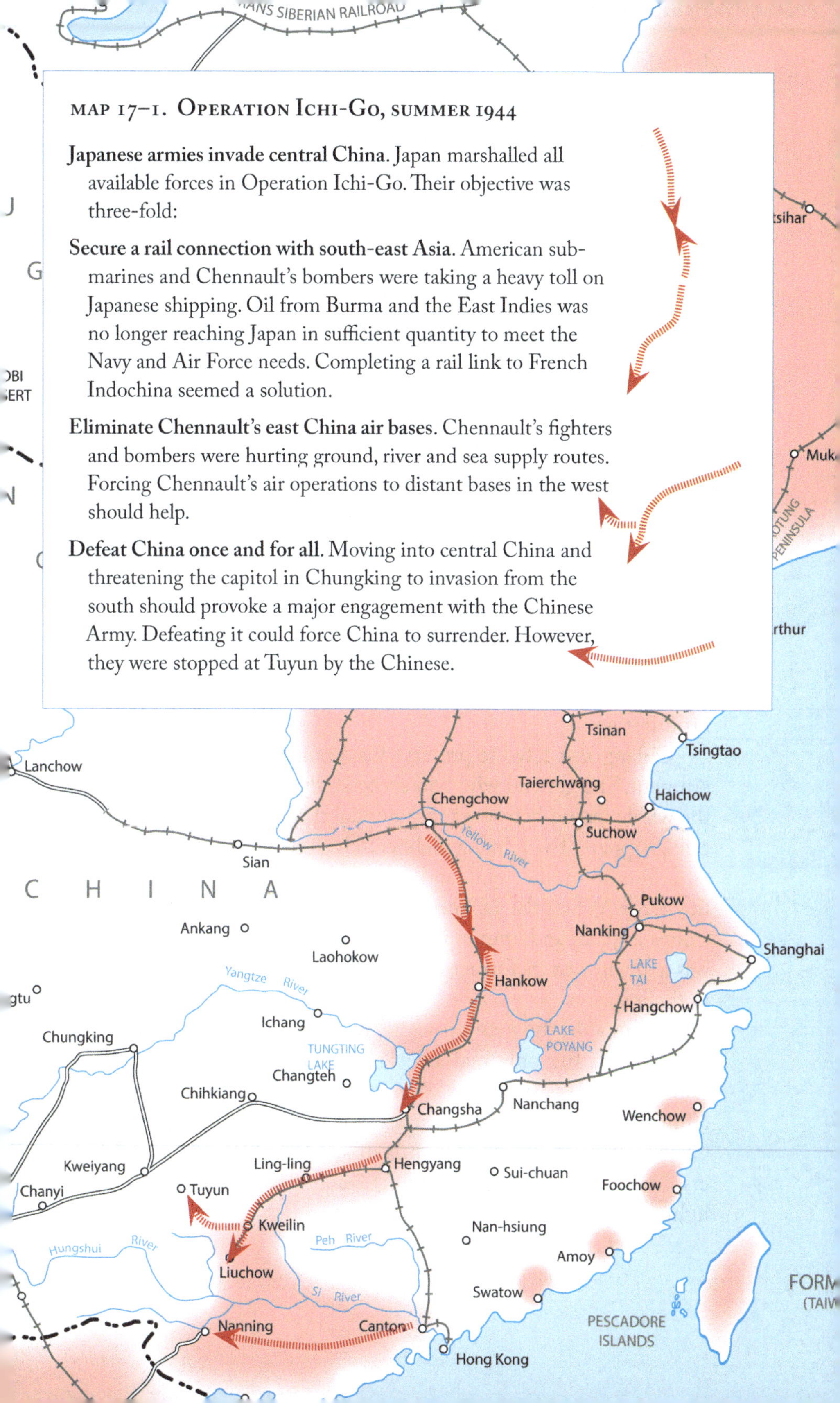

MAP 17–1. OPERATION ICHI-GO, SUMMER 1944

Japanese armies invade central China. Japan marshalled all available forces in Operation Ichi-Go. Their objective was three-fold:

Secure a rail connection with south-east Asia. American submarines and Chennault's bombers were taking a heavy toll on Japanese shipping. Oil from Burma and the East Indies was no longer reaching Japan in sufficient quantity to meet the Navy and Air Force needs. Completing a rail link to French Indochina seemed a solution.

Eliminate Chennault's east China air bases. Chennault's fighters and bombers were hurting ground, river and sea supply routes. Forcing Chennault's air operations to distant bases in the west should help.

Defeat China once and for all. Moving into central China and threatening the capitol in Chungking to invasion from the south should provoke a major engagement with the Chinese Army. Defeating it could force China to surrender. However, they were stopped at Tuyun by the Chinese.

ditional military aid from the United States, he would not fight. The United States military led by Stilwell and Bissell refused additional military aid until the Chinese fought. This was the original Catch 22: one thing must be done before another thing happens. Neither side compromised. Chiang would not provide additional troops so the United States provided no additional aid. It was quite simple, no fight-no supply.

The Allies depended on the "Stilwell memo" for guidance of their China policy. An unfortunate decision produced a string of unintended consequences for China and the United States. American policy shifted slowly away from support of the Chinese. Stilwell planted the poison; all that remained was to bite the apple.

Chiang was incredibly perceptive to recognize this change in American attitude. Flawed information in the Stilwell memo resulted in General Marshall misjudging the intentions of China. It remained so for the remainder of the war and afterwards. This predated the computer but the principle remains the same, garbage in-garbage out. Faulty intelligence confuses fact for opinion. General Stilwell forwarded his tainted opinion, not reality as it existed. This was not his finest hour.

Chiang also acted to preserve his dominance of the Chinese government. The general who had successfully defended Changsha and the whole region before was a potential threat if he became too powerful. Apparently, Chiang deliberately limited lend-lease supplies of ammunition and food, and money for provisioning troops, from this army. When the Japanese invasion struck, morale was very low, troops unpaid, and ammunition scarce. The Chinese defensive army was a shadow of its former effectiveness, and it gave little resistance to the invasion.

Chennault, the man on scene, was forced to fight with resources on hand. The only real asset available to him was the 14th Air Force. By necessity that is what he used. Here was a unique situation in the annals of air warfare and required a unique solution. Chennault's command was a mix of sundry units made up of fighters, bombers (heavy and medium) recon aircraft, cargo planes and light utility aircraft.

FIGURE 17–1. RIGHT TOOL, WRONG PLACE: B-29 SUPERFORTRESSES. Shown here at Chengtu, these huge planes were prioritized by General Hap Arnold for bombing Japan as Operation Matterhorn. Each mission needed large quantities of fuel and ordinance, requiring many supply flights over the Hump from India to stockpile enough for a single mission to Japan. Gasoline supplies to the 14th Air Force were diverted to the B-29s, hurting the 14th's effectiveness against the Japanese in China. In the end, the cost in men and materials was too high to sustain. This problem was only solved when General Curtis E. LeMay moved the B-29s to the newly captured islands of the Marianas, where they could be supplied by cargo ship. (US ARMY, 66373AC)

The 800-pound gorilla

Once again, outside events hampered the effectiveness of Chennault's Air Force. B-29 long-range bombers were coming off assembly lines, and the question for the American Joint Chiefs of Staff was where to use them. The British isles had not fallen to Hitler's invasion, so B-17 and B-24 bombers based there had the range to reach deep into Germany. The B-29 was not needed there, so that left China. Operation Matterhorn envisioned a large B-29 fleet based in safe areas of Szechuan near Chengtu. The Chinese dutifully started building four massive air bases with 8500-foot runways and infrastructure to match. That part of the operation was successful.

These bombers were designed to fly at higher altitudes than Japanese fighters or anti-aircraft artillery could reach. Starting in June 1944, six months of missions showed this to be true. However, bombing accuracy from such high altitude over the targets in Japan and Manchuria was very poor. No one understood the phenomenon of jet stream winds at the time. Releasing a bomb based on a 20 mph error in airspeed resulted in a half-mile miss. Major General Curtis LeMay arrived from Europe in August to improve things. He had led the successful strategic bombing of Germany, but here in China the logistical and operational problems with the operation were fundamental and proved insurmountable.

Planners thought that these huge aircraft could ferry their own supplies of bombs and fuel across the Hump, but this proved an utter pipe-dream. In reality, six round-trips to India were required just to stockpile enough fuel for one mission against Japan. As a result of the flying time, the resulting maintenance, and accumulated weather delays, only one mission to Japan per month proved possible. Washington viewed the B-29 program as of supreme strategic importance, and so decreed that some of the Hump cargo capacity for gasoline be diverted to Operation Matterhorn. Gasoline deliveries to the 14th Air Force dwindled starting in May.

The shortage of gasoline for the 14th Air Force practically grounded its bombers during the Ichi-Go offensive. Most of what they got went to fighters operating on ground support and railroad and river boat interdiction, and even that was limiting at times. This did not bother Chennault, he would stand on an embankment and throw rocks at the Japanese if that was the only way he could hurt them. Circumstances set the scene, his airmen made history by waging continuing air war against ground forces.

In effect Chennault was forced to fight a guerrilla war in eastern China. Lacking adequate supplies to conduct conventional extended operations, Chennault instead launched lightning raids in the vein of John Mosby. Hit fast, strike hard, and evaporate before your enemy. The Japanese were wounded and confused. Until sufficient supplies, aircraft and men arrived, this remained Chennault's chosen tactic. He became very good as a guerrilla leader of the air. It seemed to fit his fighting personality.

Battle order for the 14th Air Force

The 14th Air Force fought across a broad front. Commanders of the 308th Bomb Group and the 23rd Fighter Group waged war as best they could. Relying on innovative strategies and exceptional command abilities, they fought with limited supplies and success required a lot of luck. The premise that infantry must fight infantry in a ground offensive was correct. It was (and still is) a widely accepted concept. This was proven daily in east China.

Ground can not be held with air power alone. It's like stacking bricks without mortar: they will sit there until they are moved. Chennault had no infantry support working within his command. There was no way to deny the Japanese Army additional territory if they wanted it.

Chennault was forced to retreat from airfield to airfield. His order of the day, "Hold on until the last possible minute, then withdraw." It was always the same game, play for time. When the Japanese got too close Chennault ordered a fallback.

> Luke Williamson flew me down to Liuchow early in November for a final look at the east. Heavy clouds blanketed the east, and we slipped into Liuchow under an 800 foot ceiling in heavy rain. The Japanese were less than 50 miles away. It looked as though the weather might allow the Japanese the final triumph of catching our planes on the ground at Liuchow might not be able to fly. On the night of November 7, Casey Vincent's men began putting the torch to our base at Liuchow. The next day, with the ceiling still sagging down below the surrounding peaks, the flag was hauled down, and the planes took off for the last time. The last plane off was a silver P-51 flown by Casey, replacing a sick fighter pilot who had to be evacuated by transport. Weather was so bad that three fighter planes spun in and crashed before they reached west China bases. That night the Japanese cut the rail line 30 miles to the west of Liuchow and three days later occupied the city. (Chennault, *Way of Fighter*, 308)

Air strikes only stung the enemy. With each new attack, the Japanese grew more angry. Sometimes holding actions caught enemy columns in the open. Then there was hell to pay. No matter how effective air attacks were, the Japanese kept coming. Each delaying action bought his flyers a little more time. Air units lost one base after another. Chennault's only advantage was geography. China is a very big country. He would play this game many times in the months ahead.

The Japanese would not control eastern China any time soon. It was just too large. Anyone foolish enough to attack China via land better remember this truth.

Chennault traded space for time. The Japanese didn't have this luxury. The enemy was on a tight schedule again, with no extra time allotted. Evidence brought to Chennault by his spies indicated Japan had launched their largest and most ambitious military operation to date. Japanese ground forces received additional support from occupation troops stationed in Manchuria. This set off alarm bells all over China. Additional units from Manchuria intimated a serious offensive had begun. This was the first time these troops supported Japanese actions in China.

All signs indicated this new campaign would have far-reaching objectives. If successful, the Japanese would have a protected overland supply route across China. This was one objective the Japanese desperately needed. Their railroad would go from Peking to Indochina, no shipping needed.

Success removed the threat of Chennault's East China bases. Without these critical eastern airfields, Chennault could not defend MacArthur's operations in the Philippines or protect the left flank of Halsey's mobile carrier fleet. Because of the extended range Chennault was no longer close enough to continue his maritime air campaign against Japanese shipping.

With Japan's success in the east, it seemed an important strategic victory was being won. The Allies could no longer reach important targets throughout China and Japan. The Japanese used a good plan. Prosecuted quickly enough, the offensive might eliminate all Chinese armies in the east. There was a good chance for a Japanese victory or a military settlement with Chiang Kai-shek. If this happened, the Allied position in Asia was in great peril.

A new front could cause the downfall of Chungking and the Nationalists. Substantial Japanese troops were now engaged in a slow, relentless assault. The negative effects were it consumed large numbers of troops and critical supplies. The further Japan advanced, the more difficult resupply became. It was a case of damned if you do and damned if you don't.

Japan lost hundreds of thousands of men (killed or wounded) and tremendous expenditure of valuable supplies conducting this year-long offensive. As they slowly advanced into the heartland, problems worsened. Pressed on all fronts, Japan could ill afford these losses.

The ground campaign slowly, inevitably recorded victory after victory. Each city along their route fell. Casualties were terrible for Chinese and Japanese alike. Progress slowed because of incessant air attacks from the Americans.

This was dangerous ground for the Japanese, but success here might lead to Chinese surrender. A risk for sure, but the potential for total victory compelled them to fight on. Each limited success forced the Japanese deeper into the mainland. The Japanese strategy still demanded a quick victory. Their forces appeared on the verge of success. The High Command in Tokyo believed China could still be defeated.

Undue reliance on rigid timetables worked against the Japanese. Reality failed them at the last moment. Total victory eluded them. Japan's timeline for victory was never achieved. Asian dominance required rapid conquest. This was not going to happen and Tokyo now knew it.

Here is a classic example of the tar ball effect. Once you put your foot in, you are stuck. Like quicksand, the more you struggle the worse your situation becomes. The further the Japanese were sucked into China the less chance they would win. Success cost Japan time, supplies and manpower. These items were critical to other campaigns then being waged in the Pacific.

Warfare in eastern China was set. Japanese forces were well led and their soldiers were veteran troops. Their strategy required movement on a broad front. Japan intended to conquer almost half of China's landmass. All transportation networks and airfields in the east could be captured. Japanese field commanders planned new attacks designed to bleed the Chinese Army white. Victory here consolidated northern China as well. Isolated Chinese armies in the north would have no means of resupply. Then China must surrender. At least that was the hope of the General Staff in Tokyo.

The operation was a gambler choice: win everything or lose the war. All that stood between the Japanese and victory was the 14th Air Force and Claire Chennault. He could only launch harassing attacks. Continuous strafing against the Japanese was effective as far as it went. Chennault's instructions to all 14th Air Force commanders were simple, "if it moves, kill it."

That is exactly what they did. Chennault's forces tormented Japanese ground troops like a swarm of hungry mosquitoes. Their favorite and most deadly method for bombing troops in the open involved para-fragging attacks (low-flying bomb runs using parachuted cylinders of explosives). It brought terror and death to anyone unfortunate enough to be underneath. Para-frags exploded above the ground, showering everything below with thousands of small pieces of lethal shrapnel. Each detonation delivered destruction over a broad area. The effect is similar to being hit by a shotgun with a large killing pattern. There is very little anyone underneath can do to escape.

Death rained from the sky. Dropped from low level B-25 bombers, these attacks were most effective. Destruction of equipment and personnel was devastating. In China, para-fragging was most feared by Japanese ground forces. Infantrymen lived in constant terror. Low-level bomb runs were effective against troops, airfields and docking facilities. Para-frags caused extensive damage, and they were psychologically effective as well.

High-level strategic bombing was carried out by B-24s. In addition, both types of bombers used unconventional tactics against Japanese shipping as well. Chennault devised a new method of assault against Japanese warships and civilian transports. Heavy and medium bomber aircraft perfected skip bombing techniques for low-level attack. Flying just above the water, conventional bombs bounced off the waves and into ship hulls. They were accurate and very efficient. Chennault's tactical genius reappeared here. His maritime attacks severely damaged Japan's ability to support their forces in China.

Japan could no longer maintain their China armies. Ships could not get through Chennault's net. Teamed with American submarines, everything floating that was Japanese was swept from the sea. Strafing attacks flew regularly against maritime shipping as well. Low-flying aircraft used machine gun fire against shipping and personnel. Mariners feared these attacks. They were very effective and terrorized any survivors.

The air war continued day after day while the weather held. For Chennault's pilots, all days were alike. Locate the enemy, strafe his positions, destroy everything, and leave no one alive. When the enemy was found, he suffered. Sometimes he suffered terribly. It was like shooting ducks with a cannon. Lots of noise and no one standing after the pass.

Don Lopez remembered his first days in China with a sense of wonderment and surprise. He was excited to be flying for the famous General Chennault. Everything he saw was so different from everything he knew. He might as well be on the moon or watching Martians working around him.

> After landing, we taxied to the ATC area and parked. We gathered our gear and stepped, for the first time, onto the soil of China. My ideas about China had come from reading books, mostly by missionaries, about our friends the Chinese and from the film version of Pearl Buck's book The Good Earth. They were pretty much on the mark. The peasants looked and dressed much as they did in the film, in loose over-blouses, loose trousers, straw coolie hats, and sandals. Chinese workers were everywhere, repairing the runway, taxiways, and roads with crushed stone and mud. The stones, broken by women using small hammers, were carried off by men in baskets on the ends of yo-yo poles. Despite the hard, tedious labor, they always seemed to be in good spirits and invariably greeted the Americans with a grin, a thumbs up, and Ding Hao, Chinese for very good. Even while working in mud and dirt, they always managed to look clean and energetic. (Lopez, *Teeth of Tiger*, 60)

From the start of their tour until the last day, Americans in China found the locals mostly friendly and appreciative. Chennault insisted airmen under his command treat Chinese as worthy allies and friends. He went to great lengths to ensure this happened. The Chinese understood these men came from the other side of the world to help. Both groups believed the effort was a worthy cause. Second Lieutenant Don Lopez recalled his first action this way:

> Our strafing missions were to knock out boat and road traffic in the area east of the Yangtze River and Tung T'ing Lake, where there was a great deal of Japanese activity. Except for some rumors, we did not know that the Japanese were preparing for a major drive to the south. Several times we caught truck convoys on the road and destroyed them. Many of the trucks undoubtedly carried fuel or ammunition, because they either burned or blew up in fine fashion. It was gratifying to leave a convoy in flames; it provides tangible evidence of a mission's efficiency. The convoys usually comprised ten or twenty trucks, khaki colored, with canvas topped cargo compartments. On the open road, they were defended by light machine guns and rifles. Since we usually

> made repeated passes on the same target, we often picked up a few bullet holes from the ground fire. The damage was usually superficial, unless the coolant or oil system was hit. The coolant was the most vulnerable, since it had much more area. Even the small caliber Japanese rifles could shoot down a fighter with one hit in the coolant system. (Lopez, *Teeth of Tiger*, 92).

This situation caused great concern for Chennault's pilots who flew pass after pass against Japanese ground fire. That unlikely shot became known as the "Golden BB." This remains a common risk on low level strikes even today. It only takes one ding in the right place to cause a fighter/attack pilot to have a very bad day.

Wiltz Segura had his own experience with a golden BB as he later recalled what happened to him.

> It was my birthday, and I was on my sixty-fourth mission in China that day. I was leading the flight. Vernon Tanner was leading the other flight. We were strafing troops at Hengyang when I went down pretty low.... As I went, I heard pop, pop of their ammo across my fuselage. My wingman said, "You're hit!" and I said, "I know I am."...I was so low. They had some heavy guns on the river and somebody called out for me to be careful and not get over that position. But that was the closest way home. I didn't know how long my airplane was going to last. The minute I pulled up, I had to turn towards them. I went over about a thousand feet, trying to get enough altitude so if the engine quit, I'd bail out. Managed to get up at about three thousand feet. By this time, I knew that they had hit my oil line and it was smoking. My oil pressure was fluctuating...At three thousand feet and about ten miles from where I got hit, seven minutes later, the engine quit. I knew I was going to have to bail out-everybody knew it. They were talking on the radio. I figured it was time to get unstrapped and told the guys I was going to bail out...My buddies were flying all around until I hit the ground. The plane went off about a half a mile and crashed. They made a couple of passes around and didn't see any enemy activity. (Cornelius and Short, *Ding Hao*, 439-440)

Most casualties in the 14th Air Force came from low-level attacks, not dogfighting. "Oh, the humiliation of it," my fighter pilot friends used to say. In a fighter pilot's world, altitude translates to good and its lack is bad. Many good men and great pilots have had terrible, terrible, bad, bad days because of this little known phenomena.

Wiltz Segura found out for himself. Alone behind enemy lines, Japanese ground forces hunted him like a rabbit in a thatch patch. To an old country boy, this was not a good thing. Chennault briefed all of his men about support they could expect from local Chinese or guerilla forces should they go down. Segura found himself in the unenviable position of being shot down in Indian country.

He ran as any good rabbit would. As far and as fast as his skinny legs could carry him. Speed and endurance acquired around the track in high school earned him the nickname the Flash. After several miles of running, Segura took stock of his options. Being taken prisoner by recently strafed Japanese troops was not one of them.

> Local troops disguised as civilians were operating behind enemy lines. Officials, instead of evacuating the area, went incognito into the countryside to become guerrillas. They did tremendous work behind the lines retrieving us. I saw this farmer in a rice paddy. I was black and muddy. He saw me and took off. I decided I had better get to him before he got to the enemy, so I caught up with him. He headed for this little village of two or three huts. I got hold of him, showing him my little book which said, I'm a friend. I am American. Where's the enemy? He was scared. He led me to this little compound and then things really started happening. Guys that knew what they were doing and had guns. They came out of the hills and hid me in a little shack. They were hollering outside, giving orders. I couldn't tell if they were Japs or Chinese...I got hold of one guy and all he could say was, do not fear. They got hold of me and started running, trotting down a trail towards the woods. About two hours later we came upon some troops. (Cornelius and Short, *Ding Hao*, 441)

During the war, the most prestigious action any Chinese could do was safely recover a downed American airman. There was always fierce competition between opposing Chinese units to claim this prize.

> There, a big argument developed, and I couldn't figure out what happened until we got back about two weeks later. This was another guerrilla force. It was a great honor for a guerrilla chief to rescue an American, and when he heard that an American had gone down, he sent his men looking for me for the honor of capturing me. Chiang Kai-shek rewarded them later. We were told to let them have anything they wanted. But they wouldn't take money or anything, even food. The government apparently rewarded them very well. And I was finally returned to base. (Cornelius and Short, *Ding Hao*, 441)

While assigned to the 23rd Fighter Group, Segura almost became a double ace. He shot down eight confirmed Japanese aircraft while losing four P-40s (2 crashed, 2 shot down). If he stayed another week, he might lose one more plane, thus becoming a double ace. Holders of that distinction must shoot down five of theirs and lose five of ours.

(i.e. a double ace.) It was with a sigh of relief the aircraft acquisition officer saw Segura board a transport bound for India.

> Over the next sixty years, Segura started every birthday celebration with his original recording of "Run Rabbit, Run," a very popular tune of 1943. All things considered, Segura had a pretty good birthday. He never had another that matched the excitement. Until his dying day, Wiltz always loved the Chinese people. He remembered them with gratitude. Their efforts saved his life. (Perry Segura, *Conversation with author*, 2019)

Down the pilots came, flying out of the sun just like Chennault trained them. Down into the dangerous airspace between the ground and a merciless hail of anti-aircraft gunfire. Hoping, praying your airplane does not get hit. Chennault's men always came to kill and sometimes to die. This was the brutal experience of war. Intense and personal.

Pilots flying so low to the ground can identify individual soldiers, see their faces, shoot. If you looked, you could see the horrible results inflicted. Do not look too long, the sights are terrible, engrossing. Must not get target fixation. Get up, get altitude, get out. Then back around and do it all over again. What a war, hard and brutal, much like getting kicked in the gut by a mule. Best keep it impersonal and pretend it doesn't bother you at all.

Retreat from the east coast bases

The 14th Air Force retreated from the East China bases as the Japanese advanced. Even the main base at Kweilin was abandoned. Chennault conducted a masterful fighting retreat. Careful utilization of meager forces allowed him to slow the Japanese advance while his pilots inflicted maximum casualties. Operations by the 14th Air Force could not stop the ground offensive. It did, however, pin down the Japanese Air Force in east China. Not a bad day's work against a superior enemy. Chennault suffered minimal losses to his own forces. Units from the 14th Air Force conducted 747 sorties by four fighter squadrons and dropped 110 tons of bombs. Chennault lost only thirteen P-51s to ground fire, mechanical failure or adverse weather. This was genius, pure genius.

In November, Chinese troops stopped a Japanese thrust towards Tuyun threatening Chungking. The airbase at Chihkiang that Chennault personally started (see Chapter 1) was in a mountainous region south of Chungking. It was remote from the railroad and the thrust of the Japanese invasion, and so was not invaded or captured. Chihchiang became the major forward base remaining, and Chennault was glad he had selected its location. From there, Chennault's forces could hinder supplies moving south from Hankow. In April 1945, the Japanese did attempt to capture it, but battle hardened and well-supplied Chinese troops were airlifted in and soundly defeated the invasion. As the B-29 operation in China wound down and the Hump cargo capacity increased, gasoline supplies to the 14th resumed and more strikes on Japanese supplies were flown.

Hankow raid

The last westward thrust of the Ichi-Go offensive was stopped at Tuyun in late November of 1944. Meanwhile, the 14th Air Force ground attack operations had crippled the distribution of supplies to Ichi-Go troops, so that a substantial backlog accumulated in the port of Hankow along the Yangtze River. Chennault's intelligence network noted the Hankow buildup, and he convinced LeMay to combine forces in a major raid on Hankow.

The successful invasion of the Marianas Islands in the Pacific Ocean gave an alternative launching base and put more of the Japanese Island within B-29 range. Supplying the bombers by ship was much easier than flying the Hump. At the end of 1944, LeMay decided to move the entire B-29 operation to the Marianas bases. Chennault had been advocating using the B-29 bombers against targets in China, but the bombers had been reserved only for strategic use towards Japan. With that off the table, now was the time and LeMay agreed.

LeMay had been using incendiary bombs against German cities, as Chennault had advocated for years. Chennault planned the attack carefully. On December 18, his assembled 14th Air Force fighters and bombers attacked Hankow in timed waves, targeting the antiaircraft batteries and airfields. Just when they were done, the B-29 bombers arrived at flying at medium altitude where bombing was precise. Flying without opposition and loaded with incendiary bombs, they set the city and docks afire. Like Mandalay, it burned for days. Like

Figure 17–2. Kweilin Air Base. Photo was taken during the evacuation before the Japanese Ichi-Go offensive seized the area. Caves dug into the unique hills provided secure air raid shelters, entire repair shops and supply depots for Chennault's main forward base of operations. (US Army, 54038AC).

in Japan, these were wood and paper houses that would readily burn when attacked with fire starting weapons. LeMay and the bombers went on to the Marianas to use incendiary tactics against Japanese cities.

> The December 18 attack of the Superforts was the first mass fire-bomb raid they attempted. LeMay was thoroughly impressed by the results of this weapon against the Asiatic city. When he moved on to command the entire B-29 attack on Japan from the Marianas, LeMay switched from high altitude daylight attacks with high explosives to the devastating mass fire bomb night raids that burned the guts out of Japan… (Chennault, *Way of a Fighter*, 330)

This was the start of what became known euphemistically as the LeMay Treatment. The burn potential which Chennault and Billy McDonald originally noted in 1937 proved accurate. Japan came to fear these destructive fire raids until the end of the war.

The Japanese began retreating in December 1944, strangled by supply issues. Some of the air bases were reoccupied, and 14th Air Force operations resumed with a vengeance. Chennault had reason to be pleased with the contributions his 14th Air Force had made to the war effort.

Ichi-Go aftermath

By the time the Ichi-Go offensive was done, the strategic importance of China had shifted. Outside circumstances determined China was no longer the focus for Allied victory in Asia. Amphibious American actions against Japan's island defense chain and successful air operations by the United States Navy allowed the Allies to strike at will anywhere desired. Possession of eastern China airfields was no longer vital in Allied planning. At the beginning of 1945, the Marianas became the center for strategic bombing operations against Japan.

Essentially, the American Navy turned Japan's left flank. Allied forces no longer needed China as a base of operations. China became a prison for the trapped Japanese. Just as Chiang and Chennault had foreseen years before, the vast expanse of China and determined resistance prevented Japanese victory. There was no longer a need to fight them, just hold them in place. Japanese resources for other operations in the Pacific Theater were wasted in China.

Although initially successful, Operation Ichi-Go produced massive Japanese losses and expended vital supplies and equipment. Japan could not replace these resources. Allied operations elsewhere rendered Ichi-Go wasteful and meaningless. Japan suffered over 300,000 casualties during the twelve month offensive for no tactical or strategic gain.

Massive long range B-29 bombers flying from the Marianas reached all Japanese possessions. The success of the XXI Bomber Command of the 20th Air Force subsequently became the subject of much controversy. Many thousands of Japanese civilians were killed in these "fire raids." Destruction of this magnitude incurred indiscriminate killing of large numbers of civilians (men, women and children). Some in America thought these actions were unnecessary. Fire bombings killed innocent noncombatants. These people should remember General Grant's maxim, "War is hell." Here was the culmination of earlier Japanese actions. As ye sow, so shall ye reap. There is no room for sentiment in warfare.

Coordination between Allied naval and land-based assets was now in place. Massive raids of a thousand planes launched at will. Strongholds in the Pacific or the Home Islands faced continuous attack. Japan could no longer defend herself, but still refused to surrender.

Mobile carrier task forces ensured the Navy could strike anywhere they wished. Death tolls in these raids were horrific. Even with greatly elevated casualties, the Japanese still refused to surrender. Allied bombers continuously pounded Japan with virtual impunity.

Carrier fleets flanked the Japanese defensive ring and opened the way to Japan itself. From this point onward, China became a holding action, a military backwater, a massive prison for trapped Japanese. China's vast expanse immobilized Japan's troops. There was no immediate necessity to defeat these forces in situ.

In tactical and strategic terms, Japan gained nothing of value by isolating East China. They suffered massive casualties during the Ichi-Go offensive. Critical manpower and valuable material resources were wasted. The entire Japanese effort was well conceived, brilliantly conducted, but achieved little of import when considered in the larger picture of the war. Much like Robert E. Lee discovered on the third day at Gettysburg, he squandered precious resources (Pickett's Charge) in a magnificent but futile gesture.

Japan lost their finest field army in China by fighting in the wrong place at the wrong time. One shudders to think how those men could have been better used.

Units under his command lost 500 planes from all causes of combat while destroying 2,600 confirmed enemy planes with an additional 1,500 more listed as probable. Chennault's aircraft sank or damaged at least 2,300,000 tons of enemy merchant shipping, 44 naval vessels, 13,000 river boats under 100 tons, killed 66,700 enemy troops and knocked out 573 bridges. All this destruction happened while fighting an enemy deploying superior numbers against his small forces. It was an impressive accomplishment not duplicated in the Second World War.

> Its best testimonial came from the man who suffered the worst at our hands-Lieutenant General Takahashi, commander of the Japanese forces in central China. After the war Takahashi said "Considering all the difficulties my armies encountered in China, including guerillas, ground armies, lack of supply, difficult terrain, non-cooperation of the Chinese, I judged the operations of the Fourteenth Air Force to have constituted between 60 percent and 75 percent of our effective opposition in China. Without the air force we could have gone anywhere we wished." (Chennault, *Way of a Fighter*, 355)

There is no higher praise than a defeated enemy paying homage to the man that destroyed him. Chennault was proud of his role in the defeat of his hated enemy.

Stabbed in the back

The considered opinion of the Allied High Command once Tinian fell was that China no longer represented a major strategic priority. While Japan was clearly losing, most thought the Japanese would never surrender. Planning began for an invasion of Japan, to start in November 1945, with horrific casualties expected on both sides. An invasion in southern China was no longer needed as a pathway to Japan proper.

Chennault's presence in China was not now a necessity. In effect, Chennault's successes made him redundant. Furthermore, President Roosevelt, who had been Chennault's consistent supporter, passed away on April 12, 1945. General Marshall now had latitude to remove this constant thorn in his side. He thought maybe he would.

> With the final ebbing of the Japanese tide in China came signs that my days as American Air Force commander in China were also numbered. Attempts to oust me from my command were nothing new. The War Department had begun them early in 1942 before I even had an American command. These attempts continued in various forms ever since. (Chennault, *Way of a Fighter*, 345)

Exertion and anxiety of eight years of war took their toll on Chennault. His mind was still fit, but his body was showing its age. It was only a matter of time before Chennault reached the end of his usefulness.

Having returned to the US, Stilwell persuaded General Marshall it was mostly Chennault's fault for the failure of Stilwell's CBI war plans. All political troubles between the United States and Chiang Kai-shek resulted from Chennault's personal ill-will, Stilwell proclaimed to his old friend. Marshall was in complete agreement, as was Arnold.

> Wedemeyer read Arnold's special-delivery letter stating that it was the wish of Marshall and Arnold that the reorganization of the China air forces be carried out as planned "regardless of consequences." That phrase burned in my memory. I could hardly believe that they would commit such a bald statement to paper. But there it was. They were determined to get rid of me at all costs. Wedemeyer made it clear that he had no choice but to accede to Marshall and Arnold's wishes. Neither had the power to issue orders to Wedemeyer, but he was a junior theater commander. He owed his military eminence to Marshall and could hardly openly defy him when the issue had been made so plain in Arnold's letter. Marshall, basing his stand on Stilwell's information, was determined to remove me from any responsible post in China. (Samson, *The Flying Tiger*, 337–338)

They decided to split the 14th Air Force, and place Chennault in a backwater command. He would no longer command his fighters, bombers and men. In short, he was "offered" a significant demotion.

> On June 20, Wedemeyer told Marshall he would go along with his recommendations on the organization of his air forces and that Stratemeyer would command the China Theater air forces. Under him, Chennault would command the Strategic Air Force and Major General Howard Davidson would command the Tactical Air Force. Chennault, fed up with the whole game by

Figure 17–3. The Chennault Laugh, Rarely Photographed. His intelligence officer, Col. John O. Neal, shows a chart summarizing the accomplishments of the 14th Air Force from July 1942 to September 1944. The statistics show an impressive impact for the smallest US Air Force, operated on a shoestring. Chennault is rightly tickled. (US Army, 55085AC)

> now, tendered his request for retirement on July 6. Stratmeyer promptly approved it and appointed General Stone as Chennault's successor in command of the Fourteenth Air Force. (Samson, *The Flying Tiger*, 339)

Et tu Brute? This was the death knell for Chennault; the ring knockers finally won. They were smart, patient and they no longer needed him, so they struck. Like Brutus and Cassius and all the others who stabbed Caesar, Arnold, Marshall, Stilwell and Bissell sank their daggers deep into Chennault. Just at the moment of victory, they committed their last vile act of treachery. It was done! *Tu finis*.

Time to say goodbye

The final acts of pettiness and revenge finished Chennault. His enemies did their worst, and at the last moment, they disposed of him. There was no chance for glory going to Chennault. With all that was happening in the world, it was easy to overlook one man, no one even noticed. Chennault's enemies succeeded. He was the forgotten hero. Everything done since 1937 was overlooked and devalued. Here was the bitterest of pills to swallow.

> I flew back to Kunming with a bitter taste in my mouth. I thought of all the grim years behind me and the first bright glimmerings of victory on the horizon. (Samson, *The Flying Tiger*, 338)

There was overwhelming support from all levels of the Chinese people. They expressed genuine love for Chennault. From everyday Chinese, we see gratitude and respect. These Chinese, humble, honest, loving people cared more for him than his countrymen, his brothers in arms, his so-called American contemporaries.

From his employer, friend and confidant, he received the following,

> I have received your letter of the 17th instant and read it with much pathos. I cannot help recalling the past eight years of war at the beginning of which when China fought Japan single-handedly, you, moved by a feeling of righteous indignation, identified yourself with our cause and your devotion remained unchanged through thick and thin. You have overcome many difficulties and achieved unprecedented successes. All these facts are deeply appreciated by the Chinese nation and will always remain fresh in my memory. As I am unable to keep you here any longer, I can only comply, though very reluctantly, with your request for going home for the time being. Indeed, I cannot express in words how I am feeling at this time of bidding you farewell. However, we may part from each other, but our friendship will even grow deeper in our mind. What I wish now is that you continue to give me assistance from time to time in the expansion of the Chinese Air Force. I also believe that with your most sincere goodwill to the Chinese people and myself, you will be glad to see that my expectations come true. (Chiang Kai-shek, *letter to Chennault*, 23 July 1945).

Chennault stepped forward to aid China when no one else would. He gave everything for China, country, home and family. He did everything in his power to help save China in her darkest hour. More than anyone other than the Chinese people, Chennault stood up to aggression and tyranny while the world sat by and watched. There is no doubt in the minds of the Chinese people even today who it was that preserved China. Claire Lee Chennault found his destiny. He never hesitated once he heard China's call. China owes a tremendous debt to Chennault.

At the end, on his last day in China, this was the scene:

> In Chungking, the capital of free China, the automobile furnished him by the Generalissimo inched its way along the crowded city streets toward a public gathering place where thousands of grateful Chinese waited to present their respects to the man who had freed their lives of the terror in the sky. Along the way many Chinese and American flags flew in the breeze. Giant Flying Tiger posters and that famous Winston Churchill "V" for victory were proudly displayed. As far as he could see in all directions, Chennault was confronted with a view of a virtual sea of brown Chinese faces, bobbing and shouting and popping firecrackers, a sea which completely engulfed his car. Fearful that to drive any further might result in someone's injury, Chennault instructed the driver to kill the engine. And then it happened...As with the great exodus out of eastern China after the fall of Shanghai, as in the perilous excavation of the Burma Road, the momentum of the event was now carried by the masses. Slowly, almost reverently, the mob began to push the Generalissimo's car containing their Number One Flying Tiger, "Chen-au-duh," toward the open square. Among the cheers and jubilation, Chinese from all walks of life led Chennault to the place of honor on a wooden platform. There he received the heartfelt thanks of a nation which had survived. For the remainder of the day, countless numbers of Chinese crossed the wooden stage to shake his hand and to present him with gifts of appreciation. The scene was repeated in every city along his farewell tour. Finally, winding up in Kunming, he was told by his interpreter, "No foreigner since Marco Polo has so endeared himself to the Chinese." (Cornelius and Short, *Ding Hao*, 469-470)

Ignored by peers, belittled by contemporaries, his American superiors relegated him to the role of nuisance. He gave up everything from his previous life for China. Affection from the people of China exceeded his wildest dreams. This was China's secret, the hold they maintained on Chennault. It was his love of the masses. Here was justification for all he gave up in his previous life to come to China.

> On August 8, Chennault's C-47 rolled down the Kunming runway lined on both sides by hundreds of jubilant Chinese cheering and waving and popping firecrackers to ward off evil devils. As the plane lifted from the runway and circled the field before heading west, Chennault remembered his pledge to the Generalissimo-"I will return to China should my services ever again be needed." Thus the prophet returned home, though not without honor. His advice either ignored or unheeded, with him went hopes for a lasting peace and friendship with postwar China. (Cornelius and Short, *Ding Hao*, 470)

Chennault lived his dream. He fought for the noblest of causes. He led brave men in a righteous war. It was in China that Chennault achieved the greatest of his victories. It was as his father prophesied so long ago. Chennault's destiny lay elsewhere. He had the courage that allowed him to find and fulfill it! He would have asked for nothing more.

There is no other way to describe it. Chennault loved China and served her with nobility. China in return loved Chennault.

18

About the Author

Richard P. Voorhies Jr was born and raised in New Iberia, Louisiana. He graduated from Loyola University in New Orleans, Louisiana with degrees in English and History in 1969.

In 1970 he enlisted in the United States Air Force. After commissioning he was assigned to Undergraduate Pilot Training at Moody Air Force Base, Georgia, earning his pilot's wings in October ,1971. Second Lt.Voorhies was chosen to fly HC-130 aircraft for the Air Rescue and Recovery Service. He was stationed at RAF Woodbridge England for five years, with a follow-on assignment to McClellan AFB, California. In 1978, Captain Voorhies transferred into the United States Air Force Inactive Reserve. He joined the United States Air Force reserve in December 1980, and served in CAP-USAF Program. He additionally served as an Air Force Academy Admissions Liaison Officer from 1982 until his retirement.

Lt. Col. Voorhies is a graduate of Squadron Officers School, Air Command and Staff College and Air War College of the United States Air Force. In 1995 he retired from the Air Force Reserve.

In civilian life Voorhies worked in private industry and specialized in sales and management until his retirement in 2006. *Chennault: A Rebel in China* is the author's first book. Richard currently lives in New Orleans with his wife, Mary. Their two children and five grandchildren live near them.

Bibliography

Books

Armstrong, Alan. *Preemptive Strike: The Secret Plan That Would Have Prevented the Attack on Pearl Harbor*. Connecticut: Lyons, 2006.

Byrd, Martha. *Chennault: Giving Wings to the Tiger.* Tuscaloosa: Univ. of Alabama Press, 1987.

Bond, Charles R. and Terry H. Anderson. *A Flying Tiger's Diary*. Texas: Texas A&M, 1984.

Calloway, Nell Chennault. Author interviews (Monroe, Louisiana). Various dates.

Chennault, Anna. *A Thousand Springs: The Biography of a Marriage*. New York: Erikson, 1962.

Chennault, Claire Lee. Chennault Personal Papers. Chennault Aviation & Military Museum, Monroe, Louisiana.

Chennault, Claire Lee. "The Role of Defensive Pursuit," *Coast Artillery Journal*, 76 Nov.–Dec. 1933, 77 Jan.–Feb. 1934, 77 Mar.–Apr. 1934.

Chennault, Claire Lee. *Way of a Fighter*. Edited by Robert Hotz. New York: Putnam, 1949.

Corcoran, Thomas G. "Rendezvous with Democracy," incomplete biography.

Cornebise, Alfred Emile. *Soldier Extraordinaire: The Life and Career of Brig. Gen Frank "Pinkie" Dorn (1901–81)*. Fort Leavenworth, Kansas: Combat Studies Institute Press, 2019.

Cornelius, W. and Thayne Short. *Ding Hao: America's Air War in China, 1937–1945*. Pelican, 1980.

Crouch, Gregory. *China's Wings: War, Intrigue, Romance, and Adventure in the Middle Kingdom During the Golden Age of Flight*. New York: Bantam, 2012.

Dorn, Frank. *The Sino-Japanese War, 1937–41: From Marco Polo Bridge to Pearl Harbor*. New York: Macmillan, 1974.

Drez, Ronald J. *Predicting Pearl Harbor: Billy Mitchell and the Path to War*. Louisiana: Pelican, 2017.

Ford, Daniel. *Flying Tigers: Claire Chennault and His American Volunteers, 1941–1942*. Warbird Books, 2016.

Glines, Carrol V. *Chennault's Forgotten Warriors: The Saga of the 308th Bomb Group in China*. Pennsylvania: Schiffer, 1997.

Greenlaw, Olga S. *The Lady and the Tigers*. Massachusetts: Dutton, 1943.

Gustavsson, Håkan. *Sino-Japanese Air War: 1937–1945, The Longest Struggle*. Fronthill Media, 2016.

Hammond Contemporary World Atlas. New York: Doubleday, 1974.

Harvey, Robert. *Maverick Military Leaders: The Extraordinary Battles of Washington, Nelson, Patton, Rommel, and Others*. Vermont: Skyhorse, 2013.

Hill, David L. and Reagan Schaupp. *"Tex" Hill: Flying Tiger*. San Antonio: Universal, 2003.

Hotz, Robert B. *With General Chennault: The Story of the Flying Tigers*. New York: Coward-McCann, 1943.

Howard, James H. *Roar of the Tiger: From Flying Tigers to Mustangs, A Fighter Ace's Memoir*. New York: Crown, 1991.

Jackson, Mary Anna. *Memoirs of Stonewall Jackson*. Louisville: Prentice Press, c1895.

Kleiner, Sam. *Flying Tigers: The Untold Story of the American Pilots Who Waged a Secret War Against Japan*. New York: Viking, 2018.

Lee, Robert E. *Recollections and Letters of General Robert E. Lee*. New York: Doubleday, Page & Co., 1909.

Lopez, Donald S. *Into the Teeth of the Tiger*. Toronto; New York: Bantam, 1986.

McDonald, William C. III and Barbara Evenson. *The Shadow Tiger: Billy McDonald, Wingman to Chennault*. Birmingham: Shadow Tiger, 2016.

Mims, Sam. *Chennault of the Flying Tigers*. Philadelphia: Macrae-Smith, 1943.

Mosby, John S. *Memoirs of Colonel John S. Mosby*. Boston: Little, Brown & Co. 1917.

Neumann, Gerhard. *Herman the German*. Indiana: AuthorHouse, 2004.

Noe, Linda. Author interviews (by phone). Various dates.

Robertson, James L., Jr. *Stonewall Jackson's Book of Maxims*. Nashville: Cumberland House, 2002.

Samson, Jack. *The Flying Tiger: The True Story of General Claire Chennault and the U.S. 14th Air Force in China*. Connecticut: Lyons, 2011.

Shilling, Eric. *Destiny: A Flying Tiger's Rendezvous With Fate*. California: Ben Wal, 1997.

Schultz, Duane P. *The Maverick War: Chennault and the Flying Tigers*. New York: St. Martin's Press, 1987.

Scott, Robert L, Jr. *God is My Copilot*. New York: Scribners, 1943.

Scott, Robert L, Jr. *Flying Tiger: Chennault of China*. New York: Doubleday, 1959.

Shores, Christopher. *Air War for Burma*. London: Grub Street, 2005.

Smith, Robert T. *Tale of a Tiger*. California: Van Nuys: Tiger Originals, 1986.

Spencer, Otha C. *Flying the Hump: Memories of an Air War.* Texas: Texas A&M, 1992.

Swanston, Alexander and Malcolm. *The Historical Atlas of World War II.* New York: Chartwell, 2010.

Von Borcke, Horst. *Memoirs of The Confederate War.* Philadelphia: Lippincott, 1867.

Yu, Maochun. *The Dragon's War: Allied Operations and the Fate of China, 1937-1947.* Annapolis: Naval Institute, 2006.

Websites

US Air Force. www.af.mil/About-Us/Biographies/Display/Article/107747/clayton-lawrence-bissel/

CBI. cbi-theater.com/The Ledo Road/

HistoryNet. http://historynet.com

NARA. https://catalog.archives.gov/

Neale. www.museumofflight.org The Robert H. Neale Flying Tigers Collection/The Museum of Flight

PacificEagles. http://pacificeagles.com

Severs, Maj. Hugh G. Thesis, submitted to Air Command & Staff College, 1997. https://www.airuniversity.af.edu/ACSC/

Smith, History of the Flying Tigers, https://www.usshawkbill.com/tigers/

Ttumble Biography. https://www.usshawkbill.com/tigers/

US Army. NARA or Fold3 https://www.fold3.com/image/

Warbird Forum, ed. by Daniel Ford. http://warbirdforum.com

Warfare History Network. warfarehistorynetwork/article/operation-matterhorn/

Wikipedia. http://wikipedia.com

Index

Made in United States
Orlando, FL
04 November 2026

90215310R00164